T0326376

Integrating the Packaging and Product Experience in Food and Beverages

Related titles

Product Experience
(ISBN 978-0-08045-089-6)
Breakthrough Food Product Innovation Through Emotions Research
(ISBN 978-0-12387-712-3)
Consumer-Driven Innovation in Food and Personal Care Products
(ISBN 978-1-84569-567-5)

Woodhead Publishing Series in Food Science, Technology and Nutrition: Number 296

Integrating the Packaging and Product Experience in Food and Beverages

A Road-Map to Consumer Satisfaction

Edited by

Peter Burgess

AMSTERDAM • BOSTON • CAMBRIDGE • HEIDELBERG
LONDON • NEW YORK • OXFORD • PARIS • SAN DIEGO
SAN FRANCISCO • SINGAPORE • SYDNEY • TOKYO
Woodhead Publishing is an imprint of Elsevier

WP

WOODHEAD
PUBLISHING

Woodhead Publishing is an imprint of Elsevier
The Officers' Mess Business Centre, Royston Road, Duxford, CB22 4QH, UK
50 Hampshire Street, 5th Floor, Cambridge, MA 02139, USA
The Boulevard, Langford Lane, Kidlington, OX5 1GB, UK

Notices
Knowledge and best practice in this field are constantly changing. As new research and experience broaden our understanding, changes in research methods, professional practices, or medical treatment may become necessary.

Practitioners and researchers must always rely on their own experience and knowledge in evaluating and using any information, methods, compounds, or experiments described herein. In using such information or methods they should be mindful of their own safety and the safety of others, including parties for whom they have a professional responsibility.

To the fullest extent of the law, neither the Publisher nor the authors, contributors, or editors, assume any liability for any injury and/or damage to persons or property as a matter of products liability, negligence or otherwise, or from any use or operation of any methods, products, instructions, or ideas contained in the material herein.

British Library Cataloguing-in-Publication Data
A catalogue record for this book is available from the British Library

Library of Congress Cataloging-in-Publication Data
A catalog record for this book is available from the Library of Congress

ISBN: 978-0-08-100356-5 (print)
ISBN: 978-0-08-100360-2 (online)

For information on all Woodhead Publishing publications
visit our website at https://www.elsevier.com/

Working together
to grow libraries in
developing countries

www.elsevier.com • www.bookaid.org

Publisher: Nikki Levy
Acquisition Editor: Rob Sykes
Editorial Project Manager: Josh Bennett
Production Project Manager: Julie-Ann Stansfield
Designer: Matthew Limbert

Typeset by TNQ Books and Journals

Contents

List of Contributors   ix
Woodhead Publishing Series in Food Science, Technology and Nutrition   xi
Preface & Acknowledgments   xxv

**1 Multisensory Packaging Design: Color, Shape, Texture, Sound,
    and Smell**   1
    C. Spence
    1.1 Introduction   1
    1.2 Neuroscience-Inspired Packaging Design   2
    1.3 Packaging Color   4
    1.4 Packaging Shape   6
    1.5 Packaging Texture   9
    1.6 Packaging Weight   10
    1.7 Ease of Opening   10
    1.8 Auditory Packaging Design   12
    1.9 Olfactory Packaging Design   12
    1.10 Tasty Packaging   13
    1.11 Individual/Cultural Differences in Multisensory Packaging
         Design   13
    1.12 Conclusions   14
         References   15

2 Consumer Reactions to On-Pack Educational Messages   23
    K.G. Grunert
    2.1 The Food Label as an Information Source   23
    2.2 A Model of Consumer Processing of On-Pack Information   24
    2.3 Effects of Major Types of On-Pack Messages   26
    2.4 The Role of Context   32
    2.5 New Developments in Package Communication   33
         References   33

3 Designing Inclusive Packaging   37
    *J. Goodman-Deane, S. Waller, M. Bradley, A. Yoxall, D. Wiggins
    and P.J. Clarkson*
    3.1 Noninclusive Packaging   37
    3.2 Inclusive Design   40

3.3   A Framework for Inclusive Design   40
3.4   Empathy Tools   44
3.5   Simulation   45
3.6   Personas   52
3.7   Conclusions   54
3.8   Future Work and Trends   55
      References   56

4   Omni-Channel Retail—Challenges and Opportunities
    for Packaging Innovation   59
    C. Barnes
4.1   Introduction   59
4.2   The Omni-Channel Shopping Experience   60
4.3   Innovative Packaging for Omni-Channel Retail   65
4.4   Packaging as the Omni-Channel Integrator   72
4.5   Satisfying Customers Through Omni-Channel
      Packaging Innovation   73
4.6   Summary   74
      References   74

5   Emotion Measurements and Application to Product and Packaging
    Development   77
    S. Spinelli and M. Niedziela
5.1   Introduction   77
5.2   Emotion Measurement Methods in Sensory and Consumer
      Studies and Applied Consumer Neuroscience   78
5.3   Emotions in the Product Experience: From the Product to the
      Packaging (and Back)   99
5.4   Future Trends   109
5.5   Sources of Further Information and Advice   109
      References   110

6   Neurosense and Packaging: Understanding Consumer
    Evaluations Using Implicit Technology   121
    E. Fulcher, A. Dean and G. Trufil
6.1   Problems With the Self-Report Method in Market Research   121
6.2   Products Are Evaluated Spontaneously by Consumers   122
6.3   The Neuroscience Alternative   124
6.4   Implicit Reaction-Time Tests   125
6.5   System 1 and Associative Memory Networks   126
6.6   Implicit Versus Explicit Measures: Validity Issues   128
6.7   Cognitive Psychology of Consumer Pack Perception   129
6.8   Case Studies   130
      References   136
      Further Reading   138

7   **Explicit Methods to Capture Consumers' Responses to Packaging**   **139**

S. Thomas and M. Chambault

7.1   Introduction   **139**

7.2   Large-Scale Quantitative Assessment of Consumers' Attitudes and Perceptions of Packaging Features   **139**

7.3   Small-Scale Exploration of Consumers' Attitudes Toward Packaging Concepts and Prototypes: Focus Groups, Enabling, and Projective Techniques   **144**

7.4   Consumers' Interaction With Packaging In-home and In-store: Observation   **148**

7.5   An Evaluation of the Relative Importance of Different Packaging Components and Extrinsic Cues: Conjoint Analysis and MaxDiff   **149**

7.6   Holistic Approaches to Explore the Similarities and Differences in Overall Packaging and Packaging Design Components: Projective Mapping and Related Techniques   **153**

7.7   Conclusions   **156**

     References   **156**

8   **Consumers' Mindset: Expectations, Experience, and Satisfaction**   **161**

P. Burgess

8.1   Consumer Choices   **161**

8.2   Expectations and Satisfaction   **163**

8.3   Consumption Processing Model   **169**

8.4   Case Study: Citrus-Flavored Green Teas   **170**

8.5   Conclusions   **178**

     References   **179**

     Further Reading   **181**

Looking Forward   **183**

P. Burgess

Index   **187**

List of Contributors

C. Barnes Leeds Beckett University, Leeds, United Kingdom

M. Bradley University of Cambridge, Cambridge, United Kingdom

P. Burgess Campden BRI, Chipping Campden, United Kingdom

M. Chambault Campden BRI, Chipping Campden, United Kingdom

P.J. Clarkson University of Cambridge, Cambridge, United Kingdom

A. Dean Neurosense Ltd, Cheltenham Film Studios, Cheltenham, United Kingdom

E. Fulcher Neurosense Ltd, Cheltenham Film Studios, Cheltenham, United Kingdom

J. Goodman-Deane University of Cambridge, Cambridge, United Kingdom

K.G. Grunert Aarhus University, Aarhus, Denmark

M. Niedziela HCD Research, Flemington, NJ, United States

C. Spence Oxford University, Oxford, United Kingdom

S. Spinelli SemioSensory Research & Consulting, Prato, Italy

S. Thomas Campden BRI, Chipping Campden, United Kingdom

G. Trufil Neurosense Ltd, Cheltenham Film Studios, Cheltenham, United Kingdom

S. Waller University of Cambridge, Cambridge, United Kingdom

D. Wiggins DRW Packaging Consultants, Leicester, United Kingdom

A. Yoxall Sheffield Hallam University, Sheffield, United Kingdom

Woodhead Publishing Series in Food Science, Technology and Nutrition

1 **Chilled foods: A comprehensive guide**
   Edited by C. Dennis and M. Stringer
2 **Yoghurt: Science and technology**
   A. Y. Tamime and R. K. Robinson
3 **Food processing technology: Principles and practice**
   P. J. Fellows
4 **Bender's dictionary of nutrition and food technology Sixth edition**
   D. A. Bender
5 **Determination of veterinary residues in food**
   Edited by N. T. Crosby
6 **Food contaminants: Sources and surveillance**
   Edited by C. Creaser and R. Purchase
7 **Nitrates and nitrites in food and water**
   Edited by M. J. Hill
8 **Pesticide chemistry and bioscience: The food-environment challenge**
   Edited by G. T. Brooks and T. Roberts
9 **Pesticides: Developments, impacts and controls**
   Edited by G. A. Best and A. D. Ruthven
10 **Dietary fibre: Chemical and biological aspects**
   Edited by D. A. T. Southgate, K. W. Waldron, I. T. Johnson and G. R. Fenwick
11 **Vitamins and minerals in health and nutrition**
   M. Tolonen
12 **Technology of biscuits, crackers and cookies Second edition**
   D. Manley
13 **Instrumentation and sensors for the food industry**
   Edited by E. Kress-Rogers
14 **Food and cancer prevention: Chemical and biological aspects**
   Edited by K. W. Waldron, I. T. Johnson and G. R. Fenwick
15 **Food colloids: Proteins, lipids and polysaccharides**
   Edited by E. Dickinson and B. Bergenstahl
16 **Food emulsions and foams**
   Edited by E. Dickinson
17 **Maillard reactions in chemistry, food and health**
   Edited by T. P. Labuza, V. Monnier, J. Baynes and J. O'Brien
18 **The Maillard reaction in foods and medicine**
   Edited by J. O'Brien, H. E. Nursten, M. J. Crabbe and J. M. Ames
19 **Encapsulation and controlled release**
   Edited by D. R. Karsa and R. A. Stephenson

20   **Flavours and fragrances**
     Edited by A. D. Swift
21   **Feta and related cheeses**
     Edited by A. Y. Tamime and R. K. Robinson
22   **Biochemistry of milk products**
     Edited by A. T. Andrews and J. R. Varley
23   **Physical properties of foods and food processing systems**
     M. J. Lewis
24   **Food irradiation: A reference guide**
     V. M. Wilkinson and G. Gould
25   **Kent's technology of cereals: An introduction for students of food science and agriculture Fourth edition**
     N. L. Kent and A. D. Evers
26   **Biosensors for food analysis**
     Edited by A. O. Scott
27   **Separation processes in the food and biotechnology industries: Principles and applications**
     Edited by A. S. Grandison and M. J. Lewis
28   **Handbook of indices of food quality and authenticity**
     R. S. Singhal, P. K. Kulkarni and D. V. Rege
29   **Principles and practices for the safe processing of foods**
     D. A. Shapton and N. F. Shapton
30   **Biscuit, cookie and cracker manufacturing manuals Volume 1: Ingredients**
     D. Manley
31   **Biscuit, cookie and cracker manufacturing manuals Volume 2: Biscuit doughs**
     D. Manley
32   **Biscuit, cookie and cracker manufacturing manuals Volume 3: Biscuit dough piece forming**
     D. Manley
33   **Biscuit, cookie and cracker manufacturing manuals Volume 4: Baking and cooling of biscuits**
     D. Manley
34   **Biscuit, cookie and cracker manufacturing manuals Volume 5: Secondary processing in biscuit manufacturing**
     D. Manley
35   **Biscuit, cookie and cracker manufacturing manuals Volume 6: Biscuit packaging and storage**
     D. Manley
36   **Practical dehydration Second edition**
     M. Greensmith
37   **Lawrie's meat science Sixth edition**
     R. A. Lawrie
38   **Yoghurt: Science and technology Second edition**
     A. Y. Tamime and R. K. Robinson
39   **New ingredients in food processing: Biochemistry and agriculture**
     G. Linden and D. Lorient
40   **Benders' dictionary of nutrition and food technology Seventh edition**
     D. A. Bender and A. E. Bender

41 **Technology of biscuits, crackers and cookies Third edition**
D. Manley
42 **Food processing technology: Principles and practice Second edition**
P. J. Fellows
43 **Managing frozen foods**
Edited by C. J. Kennedy
44 **Handbook of hydrocolloids**
Edited by G. O. Phillips and P. A. Williams
45 **Food labelling**
Edited by J. R. Blanchfield
46 **Cereal biotechnology**
Edited by P. C. Morris and J. H. Bryce
47 **Food intolerance and the food industry**
Edited by T. Dean
48 **The stability and shelf-life of food**
Edited by D. Kilcast and P. Subramaniam
49 **Functional foods: Concept to product**
Edited by G. R. Gibson and C. M. Williams
50 **Chilled foods: A comprehensive guide Second edition**
Edited by M. Stringer and C. Dennis
51 **HACCP in the meat industry**
Edited by M. Brown
52 **Biscuit, cracker and cookie recipes for the food industry**
D. Manley
53 **Cereals processing technology**
Edited by G. Owens
54 **Baking problems solved**
S. P. Cauvain and L. S. Young
55 **Thermal technologies in food processing**
Edited by P. Richardson
56 **Frying: Improving quality**
Edited by J. B. Rossell
57 **Food chemical safety Volume 1: Contaminants**
Edited by D. Watson
58 **Making the most of HACCP: Learning from others' experience**
Edited by T. Mayes and S. Mortimore
59 **Food process modelling**
Edited by L. M. M. Tijskens, M. L. A. T. M. Hertog and B. M. Nicolaï
60 **EU food law: A practical guide**
Edited by K. Goodburn
61 **Extrusion cooking: Technologies and applications**
Edited by R. Guy
62 **Auditing in the food industry: From safety and quality to environmental and other audits**
Edited by M. Dillon and C. Griffith
63 **Handbook of herbs and spices Volume 1**
Edited by K. V. Peter
64 **Food product development: Maximising success**
M. Earle, R. Earle and A. Anderson

65   **Instrumentation and sensors for the food industry Second edition**
     Edited by E. Kress-Rogers and C. J. B. Brimelow
66   **Food chemical safety Volume 2: Additives**
     Edited by D. Watson
67   **Fruit and vegetable biotechnology**
     Edited by V. Valpuesta
68   **Foodborne pathogens: Hazards, risk analysis and control**
     Edited by C. de W. Blackburn and P. J. McClure
69   **Meat refrigeration**
     S. J. James and C. James
70   **Lockhart and Wiseman's crop husbandry Eighth edition**
     H. J. S. Finch, A. M. Samuel and G. P. F. Lane
71   **Safety and quality issues in fish processing**
     Edited by H. A. Bremner
72   **Minimal processing technologies in the food industries**
     Edited by T. Ohlsson and N. Bengtsson
73   **Fruit and vegetable processing: Improving quality**
     Edited by W. Jongen
74   **The nutrition handbook for food processors**
     Edited by C. J. K. Henry and C. Chapman
75   **Colour in food: Improving quality**
     Edited by D. MacDougall
76   **Meat processing: Improving quality**
     Edited by J. P. Kerry, J. F. Kerry and D. A. Ledward
77   **Microbiological risk assessment in food processing**
     Edited by M. Brown and M. Stringer
78   **Performance functional foods**
     Edited by D. Watson
79   **Functional dairy products Volume 1**
     Edited by T. Mattila-Sandholm and M. Saarela
80   **Taints and off-flavours in foods**
     Edited by B. Baigrie
81   **Yeasts in food**
     Edited by T. Boekhout and V. Robert
82   **Phytochemical functional foods**
     Edited by I. T. Johnson and G. Williamson
83   **Novel food packaging techniques**
     Edited by R. Ahvenainen
84   **Detecting pathogens in food**
     Edited by T. A. McMeekin
85   **Natural antimicrobials for the minimal processing of foods**
     Edited by S. Roller
86   **Texture in food Volume 1: Semi-solid foods**
     Edited by B. M. McKenna
87   **Dairy processing: Improving quality**
     Edited by G. Smit
88   **Hygiene in food processing: Principles and practice**
     Edited by H. L. M. Lelieveld, M. A. Mostert, B. White and J. Holah
89   **Rapid and on-line instrumentation for food quality assurance**
     Edited by I. Tothill

90 **Sausage manufacture: Principles and practice**
    E. Essien
91 **Environmentally-friendly food processing**
    Edited by B. Mattsson and U. Sonesson
92 **Bread making: Improving quality**
    Edited by S. P. Cauvain
93 **Food preservation techniques**
    Edited by P. Zeuthen and L. Bøgh-Sørensen
94 **Food authenticity and traceability**
    Edited by M. Lees
95 **Analytical methods for food additives**
    R. Wood, L. Foster, A. Damant and P. Key
96 **Handbook of herbs and spices Volume 2**
    Edited by K. V. Peter
97 **Texture in food Volume 2: Solid foods**
    Edited by D. Kilcast
98 **Proteins in food processing**
    Edited by R. Yada
99 **Detecting foreign bodies in food**
    Edited by M. Edwards
100 **Understanding and measuring the shelf-life of food**
    Edited by R. Steele
101 **Poultry meat processing and quality**
    Edited by G. Mead
102 **Functional foods, ageing and degenerative disease**
    Edited by C. Remacle and B. Reusens
103 **Mycotoxins in food: Detection and control**
    Edited by N. Magan and M. Olsen
104 **Improving the thermal processing of foods**
    Edited by P. Richardson
105 **Pesticide, veterinary and other residues in food**
    Edited by D. Watson
106 **Starch in food: Structure, functions and applications**
    Edited by A.-C. Eliasson
107 **Functional foods, cardiovascular disease and diabetes**
    Edited by A. Arnoldi
108 **Brewing: Science and practice**
    D. E. Briggs, P. A. Brookes, R. Stevens and C. A. Boulton
109 **Using cereal science and technology for the benefit of consumers: Proceedings of the 12P[th]P International ICC Cereal and Bread Congress, 24 – 26P[th]P May, 2004, Harrogate, UK**
    Edited by S. P. Cauvain, L. S. Young and S. Salmon
110 **Improving the safety of fresh meat**
    Edited by J. Sofos
111 **Understanding pathogen behaviour: Virulence, stress response and resistance**
    Edited by M. Griffiths
112 **The microwave processing of foods**
    Edited by H. Schubert and M. Regier
113 **Food safety control in the poultry industry**
    Edited by G. Mead

114 **Improving the safety of fresh fruit and vegetables**
     Edited by W. Jongen
115 **Food, diet and obesity**
     Edited by D. Mela
116 **Handbook of hygiene control in the food industry**
     Edited by H. L. M. Lelieveld, M. A. Mostert and J. Holah
117 **Detecting allergens in food**
     Edited by S. Koppelman and S. Hefle
118 **Improving the fat content of foods**
     Edited by C. Williams and J. Buttriss
119 **Improving traceability in food processing and distribution**
     Edited by I. Smith and A. Furness
120 **Flavour in food**
     Edited by A. Voilley and P. Etievant
121 **The Chorleywood bread process**
     S. P. Cauvain and L. S. Young
122 **Food spoilage microorganisms**
     Edited by C. de W. Blackburn
123 **Emerging foodborne pathogens**
     Edited by Y. Motarjemi and M. Adams
124 **Benders' dictionary of nutrition and food technology Eighth edition**
     D. A. Bender
125 **Optimising sweet taste in foods**
     Edited by W. J. Spillane
126 **Brewing: New technologies**
     Edited by C. Bamforth
127 **Handbook of herbs and spices Volume 3**
     Edited by K. V. Peter
128 **Lawrie's meat science Seventh edition**
     R. A. Lawrie in collaboration with D. A. Ledward
129 **Modifying lipids for use in food**
     Edited by F. Gunstone
130 **Meat products handbook: Practical science and technology**
     G. Feiner
131 **Food consumption and disease risk: Consumer–pathogen interactions**
     Edited by M. Potter
132 **Acrylamide and other hazardous compounds in heat-treated foods**
     Edited by K. Skog and J. Alexander
133 **Managing allergens in food**
     Edited by C. Mills, H. Wichers and K. Hoffman-Sommergruber
134 **Microbiological analysis of red meat, poultry and eggs**
     Edited by G. Mead
135 **Maximising the value of marine by-products**
     Edited by F. Shahidi
136 **Chemical migration and food contact materials**
     Edited by K. Barnes, R. Sinclair and D. Watson
137 **Understanding consumers of food products**
     Edited by L. Frewer and H. van Trijp
138 **Reducing salt in foods: Practical strategies**
     Edited by D. Kilcast and F. Angus

139   **Modelling microorganisms in food**
      Edited by S. Brul, S. Van Gerwen and M. Zwietering
140   **Tamime and Robinson's Yoghurt: Science and technology Third edition**
      A. Y. Tamime and R. K. Robinson
141   **Handbook of waste management and co-product recovery in food processing Volume 1**
      Edited by K. W. Waldron
142   **Improving the flavour of cheese**
      Edited by B. Weimer
143   **Novel food ingredients for weight control**
      Edited by C. J. K. Henry
144   **Consumer-led food product development**
      Edited by H. MacFie
145   **Functional dairy products Volume 2**
      Edited by M. Saarela
146   **Modifying flavour in food**
      Edited by A. J. Taylor and J. Hort
147   **Cheese problems solved**
      Edited by P. L. H. McSweeney
148   **Handbook of organic food safety and quality**
      Edited by J. Cooper, C. Leifert and U. Niggli
149   **Understanding and controlling the microstructure of complex foods**
      Edited by D. J. McClements
150   **Novel enzyme technology for food applications**
      Edited by R. Rastall
151   **Food preservation by pulsed electric fields: From research to application**
      Edited by H. L. M. Lelieveld and S. W. H. de Haan
152   **Technology of functional cereal products**
      Edited by B. R. Hamaker
153   **Case studies in food product development**
      Edited by M. Earle and R. Earle
154   **Delivery and controlled release of bioactives in foods and nutraceuticals**
      Edited by N. Garti
155   **Fruit and vegetable flavour: Recent advances and future prospects**
      Edited by B. Brückner and S. G. Wyllie
156   **Food fortification and supplementation: Technological, safety and regulatory aspects**
      Edited by P. Berry Ottaway
157   **Improving the health-promoting properties of fruit and vegetable products**
      Edited by F. A. Tomás-Barberán and M. I. Gil
158   **Improving seafood products for the consumer**
      Edited by T. Børresen
159   **In-pack processed foods: Improving quality**
      Edited by P. Richardson
160   **Handbook of water and energy management in food processing**
      Edited by J. Klemeš, R. Smith and J.-K. Kim
161   **Environmentally compatible food packaging**
      Edited by E. Chiellini
162   **Improving farmed fish quality and safety**
      Edited by Ø. Lie

163   **Carbohydrate-active enzymes**
Edited by K.-H. Park

164   **Chilled foods: A comprehensive guide Third edition**
Edited by M. Brown

165   **Food for the ageing population**
Edited by M. M. Raats, C. P. G. M. de Groot and W. A Van Staveren

166   **Improving the sensory and nutritional quality of fresh meat**
Edited by J. P. Kerry and D. A. Ledward

167   **Shellfish safety and quality**
Edited by S. E. Shumway and G. E. Rodrick

168   **Functional and speciality beverage technology**
Edited by P. Paquin

169   **Functional foods: Principles and technology**
M. Guo

170   **Endocrine-disrupting chemicals in food**
Edited by I. Shaw

171   **Meals in science and practice: Interdisciplinary research and business applications**
Edited by H. L. Meiselman

172   **Food constituents and oral health: Current status and future prospects**
Edited by M. Wilson

173   **Handbook of hydrocolloids Second edition**
Edited by G. O. Phillips and P. A. Williams

174   **Food processing technology: Principles and practice Third edition**
P. J. Fellows

175   **Science and technology of enrobed and filled chocolate, confectionery and bakery products**
Edited by G. Talbot

176   **Foodborne pathogens: Hazards, risk analysis and control Second edition**
Edited by C. de W. Blackburn and P. J. McClure

177   **Designing functional foods: Measuring and controlling food structure breakdown and absorption**
Edited by D. J. McClements and E. A. Decker

178   **New technologies in aquaculture: Improving production efficiency, quality and environmental management**
Edited by G. Burnell and G. Allan

179   **More baking problems solved**
S. P. Cauvain and L. S. Young

180   **Soft drink and fruit juice problems solved**
P. Ashurst and R. Hargitt

181   **Biofilms in the food and beverage industries**
Edited by P. M. Fratamico, B. A. Annous and N. W. Gunther

182   **Dairy-derived ingredients: Food and neutraceutical uses**
Edited by M. Corredig

183   **Handbook of waste management and co-product recovery in food processing Volume 2**
Edited by K. W. Waldron

184   **Innovations in food labelling**
Edited by J. Albert

185 **Delivering performance in food supply chains**
Edited by C. Mena and G. Stevens

186 **Chemical deterioration and physical instability of food and beverages**
Edited by L. H. Skibsted, J. Risbo and M. L. Andersen

187 **Managing wine quality Volume 1: Viticulture and wine quality**
Edited by A. G. Reynolds

188 **Improving the safety and quality of milk Volume 1: Milk production and processing**
Edited by M. Griffiths

189 **Improving the safety and quality of milk Volume 2: Improving quality in milk products**
Edited by M. Griffiths

190 **Cereal grains: Assessing and managing quality**
Edited by C. Wrigley and I. Batey

191 **Sensory analysis for food and beverage quality control: A practical guide**
Edited by D. Kilcast

192 **Managing wine quality Volume 2: Oenology and wine quality**
Edited by A. G. Reynolds

193 **Winemaking problems solved**
Edited by C. E. Butzke

194 **Environmental assessment and management in the food industry**
Edited by U. Sonesson, J. Berlin and F. Ziegler

195 **Consumer-driven innovation in food and personal care products**
Edited by S. R. Jaeger and H. MacFie

196 **Tracing pathogens in the food chain**
Edited by S. Brul, P. M. Fratamico and T. A. McMeekin

197 **Case studies in novel food processing technologies: Innovations in processing, packaging, and predictive modelling**
Edited by C. J. Doona, K. Kustin and F. E. Feeherry

198 **Freeze-drying of pharmaceutical and food products**
T.-C. Hua, B.-L. Liu and H. Zhang

199 **Oxidation in foods and beverages and antioxidant applications Volume 1: Understanding mechanisms of oxidation and antioxidant activity**
Edited by E. A. Decker, R. J. Elias and D. J. McClements

200 **Oxidation in foods and beverages and antioxidant applications Volume 2: Management in different industry sectors**
Edited by E. A. Decker, R. J. Elias and D. J. McClements

201 **Protective cultures, antimicrobial metabolites and bacteriophages for food and beverage biopreservation**
Edited by C. Lacroix

202 **Separation, extraction and concentration processes in the food, beverage and nutraceutical industries**
Edited by S. S. H. Rizvi

203 **Determining mycotoxins and mycotoxigenic fungi in food and feed**
Edited by S. De Saeger

204 **Developing children's food products**
Edited by D. Kilcast and F. Angus

205 **Functional foods: Concept to product Second edition**
Edited by M. Saarela

206 **Postharvest biology and technology of tropical and subtropical fruits Volume 1: Fundamental issues**
     Edited by E. M. Yahia
207 **Postharvest biology and technology of tropical and subtropical fruits Volume 2: Açai to citrus**
     Edited by E. M. Yahia
208 **Postharvest biology and technology of tropical and subtropical fruits Volume 3: Cocona to mango**
     Edited by E. M. Yahia
209 **Postharvest biology and technology of tropical and subtropical fruits Volume 4: Mangosteen to white sapote**
     Edited by E. M. Yahia
210 **Food and beverage stability and shelf life**
     Edited by D. Kilcast and P. Subramaniam
211 **Processed Meats: Improving safety, nutrition and quality**
     Edited by J. P. Kerry and J. F. Kerry
212 **Food chain integrity: A holistic approach to food traceability, safety, quality and authenticity**
     Edited by J. Hoorfar, K. Jordan, F. Butler and R. Prugger
213 **Improving the safety and quality of eggs and egg products Volume 1**
     Edited by Y. Nys, M. Bain and F. Van Immerseel
214 **Improving the safety and quality of eggs and egg products Volume 2**
     Edited by F. Van Immerseel, Y. Nys and M. Bain
215 **Animal feed contamination: Effects on livestock and food safety**
     Edited by J. Fink-Gremmels
216 **Hygienic design of food factories**
     Edited by J. Holah and H. L. M. Lelieveld
217 **Manley's technology of biscuits, crackers and cookies Fourth edition**
     Edited by D. Manley
218 **Nanotechnology in the food, beverage and nutraceutical industries**
     Edited by Q. Huang
219 **Rice quality: A guide to rice properties and analysis**
     K. R. Bhattacharya
220 **Advances in meat, poultry and seafood packaging**
     Edited by J. P. Kerry
221 **Reducing saturated fats in foods**
     Edited by G. Talbot
222 **Handbook of food proteins**
     Edited by G. O. Phillips and P. A. Williams
223 **Lifetime nutritional influences on cognition, behaviour and psychiatric illness**
     Edited by D. Benton
224 **Food machinery for the production of cereal foods, snack foods and confectionery**
     L.-M. Cheng
225 **Alcoholic beverages: Sensory evaluation and consumer research**
     Edited by J. Piggott
226 **Extrusion problems solved: Food, pet food and feed**
     M. N. Riaz and G. J. Rokey

227 **Handbook of herbs and spices Second edition Volume 1**
Edited by K. V. Peter

228 **Handbook of herbs and spices Second edition Volume 2**
Edited by K. V. Peter

229 **Breadmaking: Improving quality Second edition**
Edited by S. P. Cauvain

230 **Emerging food packaging technologies: Principles and practice**
Edited by K. L. Yam and D. S. Lee

231 **Infectious disease in aquaculture: Prevention and control**
Edited by B. Austin

232 **Diet, immunity and inflammation**
Edited by P. C. Calder and P. Yaqoob

233 **Natural food additives, ingredients and flavourings**
Edited by D. Baines and R. Seal

234 **Microbial decontamination in the food industry: Novel methods and applications**
Edited by A. Demirci and M.O. Ngadi

235 **Chemical contaminants and residues in foods**
Edited by D. Schrenk

236 **Robotics and automation in the food industry: Current and future technologies**
Edited by D. G. Caldwell

237 **Fibre-rich and wholegrain foods: Improving quality**
Edited by J. A. Delcour and K. Poutanen

238 **Computer vision technology in the food and beverage industries**
Edited by D.-W. Sun

239 **Encapsulation technologies and delivery systems for food ingredients and nutraceuticals**
Edited by N. Garti and D. J. McClements

240 **Case studies in food safety and authenticity**
Edited by J. Hoorfar

241 **Heat treatment for insect control: Developments and applications**
D. Hammond

242 **Advances in aquaculture hatchery technology**
Edited by G. Allan and G. Burnell

243 **Open innovation in the food and beverage industry**
Edited by M. Garcia Martinez

244 **Trends in packaging of food, beverages and other fast-moving consumer goods (FMCG)**
Edited by N. Farmer

245 **New analytical approaches for verifying the origin of food**
Edited by P. Brereton

246 **Microbial production of food ingredients, enzymes and nutraceuticals**
Edited by B. McNeil, D. Archer, I. Giavasis and L. Harvey

247 **Persistent organic pollutants and toxic metals in foods**
Edited by M. Rose and A. Fernandes

248 **Cereal grains for the food and beverage industries**
E. Arendt and E. Zannini

249 **Viruses in food and water: Risks, surveillance and control**
Edited by N. Cook
250 **Improving the safety and quality of nuts**
Edited by L. J. Harris
251 **Metabolomics in food and nutrition**
Edited by B. C. Weimer and C. Slupsky
252 **Food enrichment with omega-3 fatty acids**
Edited by C. Jacobsen, N. S. Nielsen, A. F. Horn and A.-D. M. Sørensen
253 **Instrumental assessment of food sensory quality: A practical guide**
Edited by D. Kilcast
254 **Food microstructures: Microscopy, measurement and modelling**
Edited by V. J. Morris and K. Groves
255 **Handbook of food powders: Processes and properties**
Edited by B. R. Bhandari, N. Bansal, M. Zhang and P. Schuck
256 **Functional ingredients from algae for foods and nutraceuticals**
Edited by H. Domínguez
257 **Satiation, satiety and the control of food intake: Theory and practice**
Edited by J. E. Blundell and F. Bellisle
258 **Hygiene in food processing: Principles and practice Second edition**
Edited by H. L. M. Lelieveld, J. Holah and D. Napper
259 **Advances in microbial food safety Volume 1**
Edited by J. Sofos
260 **Global safety of fresh produce: A handbook of best practice, innovative commercial solutions and case studies**
Edited by J. Hoorfar
261 **Human milk biochemistry and infant formula manufacturing technology**
Edited by M. Guo
262 **High throughput screening for food safety assessment: Biosensor technologies, hyperspectral imaging and practical applications**
Edited by A. K. Bhunia, M. S. Kim and C. R. Taitt
263 **Foods, nutrients and food ingredients with authorised EU health claims: Volume 1**
Edited by M. J. Sadler
264 **Handbook of food allergen detection and control**
Edited by S. Flanagan
265 **Advances in fermented foods and beverages: Improving quality, technologies and health benefits**
Edited by W. Holzapfel
266 **Metabolomics as a tool in nutrition research**
Edited by J.-L. Sébédio and L. Brennan
267 **Dietary supplements: Safety, efficacy and quality**
Edited by K. Berginc and S. Kreft
268 **Grapevine breeding programs for the wine industry**
Edited by A. G. Reynolds
269 **Handbook of antimicrobials for food safety and quality**
Edited by T. M. Taylor
270 **Managing and preventing obesity: Behavioural factors and dietary interventions**
Edited by T. P. Gill

271 Electron beam pasteurization and complementary food processing technologies
Edited by S. D. Pillai and S. Shayanfar

272 Advances in food and beverage labelling: Information and regulations
Edited by P. Berryman

273 Flavour development, analysis and perception in food and beverages
Edited by J. K. Parker, S. Elmore and L. Methven

274 Rapid sensory profiling techniques and related methods: Applications in new product development and consumer research
Edited by J. Delarue, J. B. Lawlor and M. Rogeaux

275 Advances in microbial food safety: Volume 2
Edited by J. Sofos

276 Handbook of antioxidants for food preservation
Edited by F. Shahidi

277 Lockhart and Wiseman's crop husbandry including grassland: Ninth edition
H.J. S. Finch, A. M. Samuel and G. P. F. Lane

278 Global legislation for food contact materials
Edited by J. S. Baughan

279 Colour additives for food and beverages
Edited by M. Scotter

280 A complete course in canning and related processes 14th Edition: Volume 1
Revised by S. Featherstone

281 A complete course in canning and related processes 14th Edition: Volume 2
Revised by S. Featherstone

282 A complete course in canning and related processes 14th Edition: Volume 3
Revised by S. Featherstone

283 Modifying food texture: Volume 1: Novel ingredients and processing techniques
Edited by J. Chen and A. Rosenthal

284 Modifying food texture: Volume 2: Sensory analysis, consumer requirements and preferences
Edited by J. Chen and A. Rosenthal

285 Modeling food processing operations
Edited by S. Bakalis, K. Knoerzer and P. J. Fryer

286 Foods, nutrients and food ingredients with authorised EU health claims Volume 2
Edited by M. J. Sadler

287 Feed and feeding practices in aquaculture
Edited by D. Allen Davis

288 Foodborne parasites in the food supply web: Occurrence and control
Edited by A. Gajadhar

289 Brewing microbiology: Design and technology applications for spoilage management, sensory quality and waste valorisation
Edited by A. E. Hill

290 Specialty oils and fats in food and nutrition: Properties, processing and applications
Edited by G. Talbot

291 Improving and tailoring enzymes for food quality and functionality
Edited by R. Yada

292 Emerging technologies for promoting food security: Overcoming the world food crisis
Edited by C. Madramootoo

293 **Innovation and future trends in food manufacturing and supply chain technologies**
     Edited by C. E. Leadley
294 **Functional dietary lipids: Food formulation, consumer issues and innovation for health**
     Edited by T. Sanders
295 **Handbook on natural pigments in food and beverages: Industrial applications for improving color**
     Edited by R. Carle and R. M. Schweiggert
296 **Integrating the packaging and product experience in food and beverages: A road-map to consumer satisfaction**
     Edited by P. Burgess

Preface & Acknowledgments

I recently read about a new food and beverage award scheme entitled "Het Gouden Ei"—the Golden Egg. The scheme, run by the EU-based campaign group FoodWatch, asked Dutch consumers to look back over 2015 and vote for the most misleading food or beverage product. Interestingly, they had no shortage of nominations! Although somewhat mischievous in approach, the award highlights how easy it is for a brand proposition to be eroded in the eyes of the consumer when components of the product experience, including elements such as product packaging, on-pack sensory messages, and ingredients, are not aligned with consumer expectations.

The above considerations represent the focus of this book and also have become increasingly central to the work undertaken by the Consumer & Sensory Science Department at Campden BRI.

To put this changing focus into context, the Department has over many years supported the industry by providing guidance principally in the area of food and beverage product development, and to a lesser degree associated packaging-led innovation initiatives.

Reflecting advancements in the consumer and sensory science discipline, new methods, tools, and analytical approaches have been used by the Department to facilitate optimization and product appeal among target consumers.

While these techniques have been effective at deconstructing products into their respective sensory attributes, and by linking these to hedonic measures, they have enabled products to be formulated for specific market segments, there is a growing recognition that optimizing products on this basis provides only a partial assessment of consumers' interactions with a product.

Indeed, this partial assessment carries a risk for the product developer in that perhaps additional factors such as context, usage occasion, and expectations regarding product performance could well be more significant in shaping consumers' response to the product experience than liking measures alone.

Against this background, Campden BRI has increasingly undertaken precompetitive research projects, administered under our BRI's member-funded research program, that go beyond traditional consumer and sensory science approaches. These projects include, for example, an assessment of consumers' explicit emotional responses to products, the application of evoked context approaches in product evaluation and understanding consumers' interpretations of on-pack communication, including perceptions of nutrition and health claims, among other areas.

These various diverse projects have focused the attention of the Campden BRI consumer and sensory science team on the role of packaging, its interrelationship with the product, and the impact this relationship may have on consumer opinions of the consumption experience.

As a consequence of this growing interest, a proposed new three-year member-funded project, "Packaging design - a strategic approach to enhance consumers' sensory perceptions and overall enjoyment of healthy food and drinks," received approval to progress in 2014 from Campden BRI members.

This project was designed to explore consumers' associations among design elements (eg, color, shape, and imagery) and their perceived "healthiness," in particular identifying those elements that evoke a feeling of healthiness in the consumers' mind-set.

In addition to the above-mentioned project, Campden BRI convened a seminar entitled "Consumer evaluation of packaging: techniques and applications," which was held in Chipping Campden, Gloucestershire, in October 2015.

This seminar focused on trends in the design and development of packaging, opportunities to influence consumer expectations and experiences of a product through packaging design, consumer perceptions of new packaging technologies, and approaches to measure consumer engagement with products and packaging. The seminar engaged designers, consumer and sensory scientists, and experts from psychology as well as the retail sector.

The seminar was well attended and structured in a way to encourage discussion, learning, and the sharing of experiences. Indeed, this exchange of experiences emphasized the importance of an integrated approach to packaging and product development in supporting consumer-centric innovation, particularly in light of the emerging focus in industry on the concepts of sensory branding and marketing; see, for example, Krishna, A. (2010), Sensory Marketing: research on the sensuality of products (Ed. Krishna, A.) Routledge Taylor Francis, 1–13.

The seminar was directed by Dr. Sarah Thomas (Campden BRI) with guidance and logistical support from colleagues Daphne Davies, Bethany Brown, and Phillipa Biddle.

Simon Harrop (CEO Brand Sense) expertly managed proceedings on the day, which included presentations from Peter Booth (Tin Horse), Dr. Lars Esbjerg (Aarhus University), Dr. Joy Goodman-Deane (University of Cambridge), Andrew Revell (Leeds Beckett University), Stergios Bititsios (Cambridge Design Partnership), and Dr. Eamon Fulcher (Neurosense Ltd).

This book reflects much of the content of the seminar noted above. The book effectively comprises two sections; the first four chapters focus on issues and opportunities for consideration when it comes to product and packaging design. The remaining chapters principally consider methods and techniques for assessing the integrated packaging and product experience.

In chapter "Multisensory Packaging Design: Color, Shape, Texture, Sound, and Smell," Spence critically reviews the literature on multisensory food and beverage packaging, with a focus on the contributions of the various senses to the consumer experience. Notably the influence that the multisensory attributes of the packaging can have on consumers' perception of the taste/flavor of a product is examined, with Spence concluding that interest in the multisensory attributes of product packaging is only likely to increase. Grunert in chapter "Consumer Reactions to On-Pack Educational Messages" turns the focus of the discussion to the role of the on-pack food label and the communicational elements on it. The role of "educational messages," ie, those relating to nutrition, health & nutrition claims, and sustainability messages—is considered, and the possible effects of these messages on consumers in terms of attention, understanding, inferences, and impact on choice are discussed. Goodman-Deane brings into the discussion issues surrounding inclusive design in chapter "Designing Inclusive Packaging." The key principles of inclusive design as applied to packaging are outlined and a framework for putting these principles into practice is presented. As Goodman-Deane notes, inclusive design is likely to become more of an issue in the future, given changes in the sociodemographic profile of the UK population. So while the sustainability agenda is likely to dominate packaging design, true sustainability involves the interlinking of environmental, economic, and social sustainability, of which inclusive design is a part. In chapter "Omni-Channel Retail—Challenges and Opportunities for Packaging Innovation," Barnes discusses the opportunities that omnichannel retailing presents to packaging design. As Barnes notes, in order to enhance the consumer experience in this new environment, the current rules for packaging design will need to be reassessed to optimize consumer satisfaction to reflect the shift in purchase patterns and product delivery options. The emphasis of the chapter is that, through good design and considered used of new technologies, the pack can play a key role in enhancing the consumer experience.

In chapter "Emotion Measurements and Application to Product and Packaging Development," Spinelli and Niedziela discuss the tools available in the measurement of consumer emotions elicited by a food product, from its intrinsic sensory qualities to the packaging and branding. The benefits and limitations associated with explicit, implicit, and applied consumer neuroscience tools are discussed. Fulcher continues this theme in chapter "Neurosense and Packaging: Understanding Consumer Evaluations Using Implicit Technology" by outlining the application of a specific approach, namely implicit reaction-time testing, to uncover in detail how consumers feel about a pack or brand. Thomas & Chambault in chapter "Explicit Methods to Capture Consumers' Responses to Packaging" present a series of explicit qualitative and quantitative methods to investigate consumers' perceptions, attitudes, feelings, and experiences of packaging concepts, designs, and technologies. Although these methods were presented singularly, Thomas & Chambault emphasize that many of the studies reviewed in the chapter employed a multimethod approach which emphasized the synergistic benefits derived from triangulating methods when studying phenomena of packaging from the consumer perspective.

Finally in chapter "Consumers' Mindset: Expectations, Experience, and Satisfaction," Burgess outlines a proposed approach, based on a modified version of an established scale, to assess consumption satisfaction. As the author comments, the satisfaction response is a complex construct consisting of cognitive, hedonic, and affective components. The satisfaction scale has the potential to provide the brand owner with a broader perspective on the consumption experience and therefore guidance on developing strategies that will secure a strong preference for the brand, a commitment to continued purchase and the foundation for long-term brand survival.

It is hoped that the book, by bringing together current thinking around the inter-relationship between packaging and the product and its effect on the consumption experience, will have appeal and relevance to those with interests in packaging and/or product development along with those with a general interest in the field of consumer and sensory sciences.

Additionally, it is intended that the book will further stimulate interest in improving understanding of the deep emotional influences on consumer choices and preferences and awareness and of the need to align the brand promise with the product and packaging experience. Hopefully the book will also provide a platform to encourage collaborative dialogue between brand and R&D teams, leading to greater congruency for the consumer.

So, my thanks go to the organizers, chair, speakers, and participants for making the above-mentioned seminar a success, and to the contributions from the speakers and authors for making this book possible.

A final thanks goes to the team at Woodhead Publishing, in particular to Rob Sykes and Josh Bennett, for their patience, support, and gentle coaxing of contributors and management of the chapter submissions.

Peter Burgess
Campden BRI
December 2015

Multisensory Packaging Design: Color, Shape, Texture, Sound, and Smell

1

C. Spence
Oxford University, Oxford, United Kingdom

1.1 Introduction

The multisensory attributes of the packaging undoubtedly constitute a key element in the success of many, if not all, mass market food and beverage products (eg, Klimchuk and Krasovec, 2013; Moskowitz et al., 2009; Paine and Paine, 1992). In recent years, the function of food and beverage packaging has certainly gone far beyond its original role in portion control and product preservation (see Hine, 1995 for a historical overview). In fact, as the decades have passed by, the key role(s) played by the packaging in marketing has become increasingly clear (see Calder, 1983; Day, 1985; Lannon, 1986; Masten, 1988; Pilditch, 1973; Sacharow, 1982; Schlossberg, 1990; Simms and Trott, 2010; Underwood and Ozanne, 1998; Vartan and Rosenfeld, 1987); Nowadays, then, packaging really is the fifth "P" in the marketing mix (eg, Nickels and Jolsen, 1976). Here at the outset, though, it is perhaps worth bearing in mind that the cost of the packaging far exceeds that of the contents in many product categories (see Spence and Piqueras-Fiszman, 2012). No wonder, then, that getting the packaging "right" has become such a key element of the marketing strategy for many companies when it comes to trying to ensure the long-term success of their products in the increasingly competitive marketplace. At the same time, however, many companies are having to deal with a growing consumer and governmental backlash against what is perceived (by many) to be an excess of packaging (eg, see Anon, 2006; Finch and Smithers, 2006; Usborne, 2012).

In this review, I will focus on the multisensory aspects of packaging design for food and beverage products, an area of growing interest in recent years (eg, Anon, 2010b; Day, 1985; Hruby and Sorensen, 1999; Spence and Piqueras-Fiszman, 2012). Food and beverage packaging is a particularly intriguing category because it is one in which the packaging has to serve multiple functions. On the one hand, it obviously needs to stand out on the shelf (just as for other product categories). However, given that it has been estimated that we consume as much as a third of the food products we buy direct from the packaging, which certainly needs to be optimized for the consumption experience as well. In fact, a growing body of both anecdotal and empirical research now shows that changing the multisensory design of the packaging can significantly affect people's judgments of the contents (eg, Mohan, 2013; Raine, 2007). No surprise, then, that over the last few years it has been estimated that a third of the world's largest brands have been working on "sensory branding" strategies (Johnson,

Integrating the Packaging and Product Experience in Food and Beverages. http://dx.doi.org/10.1016/B978-0-08-100356-5.00001-2

2007). Indeed, time and again, it would seem that consumers have a hard time when it comes to trying to report solely on the sensory/hedonic properties of products themselves. That is, they are often influenced in their evaluations by the extrinsic (Underwood, 1993) sensory properties of, and meanings attached to, the product packaging (see Spence and Piqueras-Fiszman, 2012). More often than not, what one sees, at least in the setting of the laboratory, is that people's feelings about the packaging tend to carry over and influence what they say about the contents (that is, the product itself) when they come to taste/evaluate them. Such effects have been described in terms of the notion of "sensation transference" (Cheskin, 1957; Spence and Piqueras-Fiszman, 2012) or "affective ventriloquism" (Spence and Gallace, 2011).

The fact that our perception of a product can be so radically affected by the multisensory design of the packaging obviously raises troubling questions concerning the utility of so much blind food and beverage testing, where products are evaluated away from any of their packaging (Davis, 1987). How can one ever hope to predict the ultimate success of a product under such conditions, one might well ask (Spence, 2009)?

The focus in this review will be on the various sensory aspects of the packaging. I want to start by looking at those attributes of the packaging that can be ascertained visually, namely the color, shape, and texture—though, of course, the latter two can also be experienced haptically in the consumer's hands (eg, Piqueras-Fiszman and Spence, 2012a; Spence and Piqueras-Fiszman, 2012). The sound made by the packaging when a product is picked up from off of the shelf, or when it is opened, also constitutes a potentially important, if often overlooked, aspect of the consumer's overall multisensory product experience (see Byron, 2012; Spence and Piqueras-Fiszman, 2012; Spence and Wang, 2015). Finally, there has been a recent growth of interest in scent-enabled packaging; and, as we will see later, some innovative souls are even considering the possible market for edible packaging (just think of the analogy with the skin of the grape; see http://www.wikipearl.com/).

1.2   Neuroscience-Inspired Packaging Design

Part of the recent growth of interest in multisensory packaging design undoubtedly stems from the potential utilization of some of the latest research techniques from the field of experimental psychology and cognitive neuroscience. Increasingly, such methods are being used to help packaging designers discriminate between the different design alternatives that they might be considering. Techniques such as the implicit association test (see Parise and Spence, 2012; Piqueras-Fiszman and Spence, 2011; see also Maison et al., 2004) and eye-tracking (Clement, 2007), especially when combined with other techniques such as word analysis (eg, Piqueras-Fiszman et al., 2013; see also Ares and Deliza, 2010) or the analysis of a consumer's grasping behavior (Juravle et al., 2015; see also Desanghere and Marotta, 2011), would seem especially promising here. That said, not all of the innovative techniques that have been tried in the area of multisensory packaging design have proved successful (eg, see Durgee and O'Connor, 1996).

Another area of rapidly growing research interest relates to the use of online testing platforms such as Mechanical Turk and Prolific Academic (see Woods et al., 2015 for a recent review). These online resources are increasingly enabling researchers to evaluate the relative merits of various different packaging designs in diverse markets at surprisingly low cost (and in a time frame that is likely to keep the marketing manager happy). In our own work in this area, for instance, we have often been able to collect data from more than 300 participants in less than an hour (eg, Velasco et al., 2015b). While the spread in terms of the participant base is still not ideal when it comes to answering many marketing questions, the explosion of online testing resources is definitely one emerging approach to packaging research to watch closely in the coming years.

In recent years, a growing number of research practitioners have become increasingly excited by the possibilities associated with the use of cognitive neuroscience techniques such as functional magnetic resonance imaging and event-related potentials. The suggestion from the neuromarketers is that such research methods can be used to evaluate the design of product packaging more directly, that is, without having to rely on what the consumer says that they are going to do, or which of a range of packaging design alternatives they indicate that they prefer (eg, Basso et al., 2014; Pradeep, 2010; Stoll et al., 2008). My suspicion, though, is that such enthusiasm for the neuromarketing approach is currently misplaced, especially when it comes to the evaluation of multisensory packaging design solutions, that is, solutions that go beyond the merely visual. (And that is before one starts to think about the consequences of countries such as France banning brain-scanning for business (Oullier, 2012).) For, at least in my experience working with industry over the years, the majority of neuroimaging techniques generally tend to be too slow and too expensive to utilize in most real-world commercial situations. (No wonder, then, that so many neuromarketing companies have started to shift more toward online behavioral testing in recent years.) What is also worth bearing in mind here is that researchers interested in packaging design have actually been using electrophysiological brain-imaging techniques for more than three decades now with, it has to be said, remarkably limited success (eg, see Weinstein, 1981 for one early example). And while it may certainly be true that incorporating colorful brain scans into the presentations was, at least for a time, an influential marketing tool in its own right (see McCabe and Castel, 2008; Weisberg et al., 2008), the latest evidence suggests that the seductive allure of such images is probably starting to wane (Michael et al., 2013; Spence, 2015). Hence, for all intents and purposes, I believe that it is the psychological sciences rather than the cognitive neurosciences that may have more to offer the packaging design agency, marketer, or brand manager when it comes to innovating in terms of their design solutions.

All that being said, I am firmly of the belief that one's study designs when it comes to evaluating packaging designs can, and probably should, be inspired by the latest insights emerging from neuroscience, hence the title of this section. And while the neuroscience-inspired approach is unlikely to offer any design solutions in its own right, it can nevertheless help to provide robust methods for discriminating between different (possibly quite similar) design alternatives. Although the transition is undoubtedly a slow, and for some a painful, one, I think we are starting to see the gradual decline of the once ubiquitous focus group in the field of packaging design (Lunt, 1981), and its replacement by neuroscience-inspired techniques. (No bad thing in my opinion!)

In part, the growing popularity of such alternative testing techniques has been necessitated by the growing realization that perception is inherently multisensory. That is, none of us, psychologists and neuroscientists included, can unpick our senses (see also Pinson, 1986). We all perceive only the integrated output of our senses (see Spence, 2009), thus meaning that none of us are particularly good at determining the actual senses that may actually be responsible when introspecting about our own perception.

In the sections that follow, I will look at each of the sensory elements of product packaging.

1.3   Packaging Color

Let's start by looking at color (Plasschaert, 1995; Salgado-Montejo et al., 2014). Color may well be the single most important sensory feature of product packaging. It should come as little surprise, then, to find that color is used by the majority of food and beverage brands in order to indicate the type/flavor of product that can be found within (Danger, 1987; Gimba, 1998). Indeed, according to an informal store audit conducted some years ago by Garber et al. (2001), more than 90% of brands on the supermarket shelf used packaging color to convey relevant information about the contents. What is clear is that in everything from carbonated beverages to pharmaceuticals, the color of the packaging sets expectations about the properties of the contents (eg, Esterl, 2011; Garber and Hyatt, 2003; Roullet and Droulers, 2005; Lynn, 1981; Wan et al., 2015). It likely also influences the consumer's purchase behavior (Seher et al., 2012). Packaging color can also be used to capture attention (Danger, 1987; Marshall et al., 2006; Sacharow, 1970) at what some have chosen to term the "First Moment of Truth" (Louw and Kimber, 2011)—this is something that is especially important once it is realized that the average shopper may see as many as 1000 different products per minute as they walk down the aisles of the average supermarket (Nancarrow et al., 1998).

One thing that is especially interesting about the use of color in product packaging is that the meaning varies by country and by product category (see Wan et al., 2014). And while different brands may use different color–flavor conventions from their competitors (in order to stand out on the supermarket shelf), this is, generally speaking, a risky strategy to adopt (Piqueras-Fiszman and Spence, 2012d). Furthermore, while the meaning of the hue of packaging color may vary, my suspicion is that the saturation/intensity likely conveys the same message regardless of category and regardless of country. Namely, the more saturated the color of the packaging, the stronger/more intense the taste/flavor of the contents is likely to be. That is, strong, bold colors in product packaging generally signify richer flavors and more intense taste experiences.

Piqueras-Fiszman and Spence (2011) conducted a study of the meaning and impact of packaging color for potato chips (or crisps, as they are called in the United Kingdom). These researchers demonstrated that when cheese-and-onion-flavored crisps were placed in the packaging of ready-salted crisps and given to unsuspecting participants to try in the laboratory, people sometimes perceived that the crisps tasted of the flavor that was associated with the packaging color, or else identified some other flavor entirely (perhaps a combination of the inherent taste/flavor of the crisp itself together with the flavor expectations set by the packaging). Intriguingly, Piqueras-Fiszman and Spence were also able to highlight the different flavor meaning of packaging color depending on a consumer's specific brand affiliation. So, for example, in the United Kingdom (not to mention in several other countries), light blue is used to signal "salt and vinegar"-flavored crisps, whereas green is associated with "cheese and onion" flavor. However, the Walkers brand has used the opposite color–flavor convention to everyone else in the marketplace since 1983. Somehow, perhaps due to their playful marketing strategy/image, Walkers has managed to succeed with this unconventional strategy for more than three decades now. Under such conditions, the meaning of the color of participants' flavor perception was shown to depend on their brand affiliations.

It is interesting to note that the impact of packaging color on product perception has also been seen in the beverage category (eg, Barnett and Spence, submitted; Cheskin, 1957; Esterl, 2011). In fact, it was Louis Cheskin who first reported that consumers rated 7-Up as tasting more lemony/limy when drinking from a can that was yellower than normal. Meanwhile, more recently, Esterl noted that many North American consumers had been complaining about the change in the taste of their Coke when a limited edition white-colored Christmas can was released.

While marketers often try to establish universal or at least culture-specific meanings, it is important to note that the meaning of color in the food and beverage category is often determined by the particular product category (or image mold) in which that color happens to be presented. Just think, for example, of the meaning of the color red: This color conventionally signifies "ready-salted" when it comes to crisps. However, one and the same color could well mean "no fat" when it comes to the milk aisle. As such, I worry that many of those marketing studies that have attempted to establish the context-independent meanings attached to colors by those in different cultures may be of limited value (eg, Jacobs et al., 1991; Madden et al., 2000; Tutssel, 2001; Wheatley, 1973). To my mind, color is nearly always seen in context.

Of course, given the increasingly global nature of the marketplace for food and beverage brands, one question of growing interest to many brand owners concerns how to choose a color for one's packaging that will have the appropriate meaning in different markets around the world (see also Lowenthal, 1981). To this end, we have recently conducted a series of cross-cultural studies utilizing the power of online testing (see above) in order to determine the putatively appropriate packaging colors for a range of novel flavors of crisps (Velasco et al., 2014b; Wan et al., 2014). For instance, the participants in one study were invited to view a number of packets advertising different flavors of crisps. The participants had to pick the most appropriate color for each flavor from a preselected palate of seven alternatives (yellow, blue, orange, fuchsia, red, green, and burgundy). The results highlighted some degree of cross-cultural consistency: So, for example, the Chinese, Colombian, and British participants whom we tested all picked green as the most appropriate color for cucumber-flavored crisps. When it came to certain other flavors, however, a number of clear cross-cultural differences in flavor–color associations were apparent. The suggestion here was that the color–flavor associations that were, in some sense, arbitrary were more likely to show cross-cultural associations.

Sometimes, of course, the color that is used for product packaging indicates not the flavor of the product itself (at least not directly) but rather represents a distinctive color associated with a specific brand. Think Coca-Cola's red, Cadbury's Dairy Milk purple, Barilla pasta's blue, or Heinz Baked Beans's bluey-green as representative examples. Given the growing popularity of open windows in product packaging, it can be argued that, looking forward, the most successful brand colors are likely going to be those that provide a good (ie, visually appealing) color contrast when the product is seen through the transparent window against the branded color of the packaging. Excellent examples here are the red of the baked beans against the Heinz brand color, the purple/brown combination of Cadbury's Dairy Milk, and the yellow of pasta against the brilliant dark blue of Barilla. (I have also seen a number of less successful examples over the years. I am not sure how many of them are still in the marketplace. Not many would be my guess.) Given the importance of color to many brands, changing that brand color is then a move to be undertaken only after careful consideration (see Anon, 2013; Cooper, 1996; Esterl, 2011).

1.4 Packaging Shape

According to some prominent marketers (eg, Lindstrom, 2005), the shape of packaging falls squarely under the heading of tactile branding. I would be tempted, however, to argue that such a view fundamentally misses the point that we nearly always look at product packaging before we pick it up (see Juravle et al., 2015). Moreover, given that we are visually dominant creatures, the seen shape of the packaging is likely to have a much greater impact on the expectations and hence on the subsequent experience of consumers than the felt shape of the packaging in their hands. Not that the feel of the packaging isn't important; It most certainly is! It is just that we normally see the color and shape of the packaging long before we feel it, and hence those visual cues anchor and dominate the subsequent experience (Piqueras-Fiszman and Spence, 2015). For those wanting to enhance the feel of their packaging, then, it is better to concentrate on its texture and/or weight (see below).

In recent years, there has undoubtedly been a great deal of innovation in terms of packaging shape (eg, Bertrand, 2002; Miller, 1994; Prone, 1993; Schlossberg, 1990). Here, though, it is important to consider the powerful notion of the "image mold" (Meyers, 1981): This is the name given to a particular packaging shape that has come to be associated in the mind of the consumer with a specific class of product, or on occasion its brand. One of the classic examples is the Wishbone salad dressing bottle. In theory at least, salad dressing could come in bottles of virtually any shape. Yet, due to the success of the Wishbone brand, this has now become the standard shape for bottles of salad dressing in the marketplace. It has, in other words, taken on the status of an image mold. One can think also of the cylindrical container as the image mold for premium ice cream (Cheskin, 1957). Even if shown nothing more than the

Figure 1.1 Silhouette of the Wishbone salad dressing bottle. This image mold conveys the notion of salad dressing in the mind of the majority of North Americans.

silhouette (see Fig. 1.1), many North American consumers are still remarkably good at identifying the product category.

Other examples of branded image molds include the Kikkoman dispenser bottle (Blythe, 2001, p. 116; Day, 1985), the Perrier water bottle, the Brahma beer bottle from Brazil, etc. (Hine, 1995; Johnson, 2007; Miller, 1994). The most powerful of all image molds, though, has to be the branded Coca-Cola contour bottle, first introduced nearly a century ago (Prince, 1994). It is interesting to note here how even as the packaging format has changed over the years, companies like Coca-Cola (and for that matter Heinz when purveying sachets of tomato ketchup) have often chosen to place a black silhouette of the classic image mold of their signature packaging on the side of their new packaging format (Durgee, 2003; see also Arboleda and Arce-Lopera, 2015). I am often struck in my discussions with industry by how many strong national, and on occasion international, brands have no image mold to speak of. I would argue that one cannot underestimate just how much of an impact serving the same food or beverage product in a different packaging format can have on the consumer's multisensory product experience (eg, see Bititsios, 2012).

Now, there is growing interest from marketers and packaging designers wishing to reposition their product by "borrowing" the image mold from an already established product in another category. By so doing, the hope is that they can acquire any positive associations that image mold may hold in the minds of their target consumers (see also Associated Press, 2013). One especially successful example from the UK marketplace in recent years has been the packaging of soup by the Covent Garden Food Co. in the Tetra Pak format (ie, rather than in cans; think Campbell's; Stern, 1981; see Fig. 1.2). Note that this packaging format was formerly associated with milk in the United Kingdom

Figure 1.2 The New Covent Garden Co.'s Tetra Pak carton successfully captures the notion of freshness and conveys the naturalness of its ingredients successfully.
Copyright the author.

(Meyers, 1981), hence it had associations with naturalness and freshness. Other players in this space currently include those packaging their high-end olive oil in what look like oversized perfume bottles. Here, though, while the association with an expensive product is certainly achieved, I do worry that the perfume bottle has strong negative associations with liquids that should most definitely not be ingested. Not exactly what one wants with a premium olive oil is my guess.

For those who do not wish to utilize a preexisting image mold for their food or beverage product, then what shape of packaging should they use (or introduce)? Here the neuroscience-inspired approach provides some insights based on the emerging literature on crossmodal correspondences. Crossmodal correspondences have been defined as the surprising cross-sensory associations that many of us share between seemingly unrelated dimensions of experience in different sensory modalities. Crossmodal correspondences have recently been established between tastes, flavors, aromas, and curvilinearity of form. In terms of assessing the shape properties that might correspond with the taste/flavor of a particular product, neuroscience-inspired methods could certainly help. So, for instance, Velasco et al. (2014a) conducted a study in which they assessed which packaging format was best associated with a hypothetical sweet- or sour-tasting product (see also Overbeeke and Peters, 1991; Smets and Overbeeke, 1995). Here, of course, one can think about not only the roundness vs angularity of the packaging itself, but also the roundness/angularity of the label/logo (Ngo et al., 2012; Westerman et al., 2013) or even the typeface (Velasco et al., 2015a). Becker et al. (2011) have shown that the curvature of the packaging affects the consumer's ratings of the taste of a yogurt, while Ares and Deliza (2010) reported that people associate rounder yogurt containers with creamier contents.

Ultimately, then, when thinking about packaging shape, there are two opposing constraints on packaging design: On the one hand, there is the powerful notion of the "image mold"; on the other, there are likely underlying crossmodal correspondences between the taste/flavor of the product and the shape of the packaging/logo (eg, Spence, 2011, 2012; Spinney, 2013; Velasco et al., 2014a). In some cases, of course, the marketers/packaging designers will intuitively have stumbled on the underlying correspondence (see Dichter, 1971), and over time that may have become incorporated into the image mold or convention for the category (see Spence, 2012).

1.5 Packaging Texture

I would say that the texture of product packaging constitutes an important, if relatively underexplored, component of the consumer's overall multisensory product experience (see Anon, 1999; Gallace and Spence, 2014; Spence and Gallace, 2011; Zampini et al., 2006). To date, the impact of variations in packaging texture has primarily been studied in the laboratory setting and in focus groups (Anon, 1999). Packaging designers may have one of a number of objectives in mind when it comes to thinking about changing the texture, or feel, of the packaging. Here, it is worth noting that the packaging designers may wish to convey a notion of a "natural" feel (eg, Labbe et al., 2013; Nikolaidou, 2011) or else to convey a certain affective response in the mind of the consumer who is handling the packaging (Chen et al., 2009; Schifferstein et al., 2013) by the feel of the packaging.

One question that researchers have only recently started to address is whether the feel of the product packaging can influence the consumer's experience of those products that are consumed direct from the packaging. Piqueras-Fiszman and Spence (2012a) addressed this question in a study in which participants were either presented with yogurt or else with pieces of digestive biscuit served in plastic yogurt pots that either had their normal smooth sides or else had been treated to give them a much rougher feel (by adhering a sheet of rough sandpaper to the outside of the packaging). The participants had to rate the texture of the food and their liking of it. The results showed that people's rating of the texture of the digestive biscuit, if not of the yogurt, was significantly affected by the feel of the packaging (rough vs smooth). Now, while no one is seriously thinking about coating their product packaging with sandpaper (as done by Piqueras-Fiszman and Spence, 2012a as just a proof-of-principle study to highlight the potential impact of the feel of the packaging on people's perception of the contents), I have come across one vodka manufacturer who wanted to indicate the strength of the alcohol by the roughness of the sandpaper on which the label was printed (see Spence and Piqueras-Fiszman, 2012)!

In a conceptually similar study, Krishna and Morrin (2008) have shown that if the feel of the container in which a drink is presented is too flimsy, it can also negatively influence how consumers rate the contents (see also Becker et al., 2011; Biggs et al., submitted; Tu et al., 2015).

Giving one's product packaging an interesting feel, or finish, may also constitute an effective marketing tool in that it may encourage the consumer to pick the product up off the shelf, and by so doing, increase the likelihood that they will end up placing the product in their basket (see Gallace and Spence, 2014; Spence and Gallace, 2011). Here, one might think of everything from the Heineken can that had been coated with tactile paint (Anon, 2010a), through to those bottles and cans with a raised crest or logo on the front-facing side. An interesting, and often distinctive, tactile/haptic feature, one that encourages the consumer to touch and/or pick-up the product, will if anything increase the likelihood of purchase. Of course, such unusual packaging features always have a cost implication attached. Unfortunately, too often it is the case that these features end up being removed from the packaging to save money. In my opinion, this is often a mistake.

1.6   Packaging Weight

One aspect of the feel of the packaging that is absolutely crucial to modulating the consumer product experience is its weight (see Piqueras-Fiszman and Spence, 2012b). Over the last few years, we have conducted a number of studies demonstrating that the consumer's perception of the sensory and hedonic properties of a range of food and beverage products can be altered significantly simply by changing the weight of the packaging in which that product happens to be presented. Generally speaking, those products that are presented in heavier packaging will be generally rated as having a more intense smell (Gatti et al., 2014), as likely to be more satiating (Piqueras-Fiszman and Spence, 2012c; Spence and Piqueras-Fiszman, 2011), and to be of better quality (Kampfer et al., submitted). It is interesting here to note that across a relatively diverse range of product categories, including everything from bottles of wine to lipstick, a strong correlation exists between the weight of the packaging and the price of the product. That being said, there are undoubtedly challenges here, especially given that, as noted earlier, many companies are increasingly being forced to reduce the weight/amount of their packaging. Nevertheless, when one sees the beneficial effects of increased packaging weight on the consumer's multisensory product experience, then the trade-off becomes all the more salient.

Intriguingly, some packaging designers are currently considering whether there are any psychological tricks that can be used to increase the perceived weight without actually adding any more weight to the packaging. Here, one might think of the fact that certain colors appear heavier than others (Alexander and Shansky, 1976). There may also be opportunities here around changing the affordance points for the grasping of packaging (see Spence and Piqueras-Fiszman, 2011).

1.7   Ease of Opening

Now, before closing this section on the haptic aspects of packaging design, it is probably worth pausing for a moment to think about how easy it is to pick up the packaging and to open and/or pour from it. While classic early research suggested that harder-to-open packaging was associated in the mind of the consumer with a higher quality product (McDaniel and Baker, 1977), the number of injuries annually that are attributable to packaging that consumers simply find too difficult to open should certainly give the packaging designer cause for concern (eg, Whyte, 2013). Here, in terms of differences in ease of opening, one could also contrast a bottle of beer with a twist-off cap versus one that requires a bottle opener (Stuckey, 2012, p. 296). The suggestion being that the difference in effort might well impact the consumer experience, whether the consumer realizes it or not.

Another aspect of ease of opening concerns the number of layers of packaging that the consumer has to work their way through in order to reach the product. There is certainly a role here for adding extra layers of packaging in order to recreate the impression of giving a present. This includes everything from the bottle of wine that comes in a presentation box/case, through to the many layers of packaging that one would traditionally find when trying to get to taste the chocolate in one's Easter egg (eg, Barthel, 1989; see also Rigby, 2010/2011).

Here, one other important issue relates to handedness: When designing asymmetrical packaging it is worth pausing to consider just how easy the average (right-handed) consumer will find it to pick up and use (eg, pour). The important point to note here is that those products that come in packages that afford grasping (and pouring) by the right-hander are likely to be at something of a competitive advantage in the marketplace relative to other products that are sold in packages that are a little more difficult to manipulate (eg, which are in some sense designed for the left-hander; see Fig. 1.3; see also Elder and Krishna, 2012). Another benefit of ease of use is, of course, the potential for increased product usage. Just take, for example, the EZ Squirt plastic ketchup bottle. This innovative packaging design (and the associated increased ease of use) apparently increased consumption by a not inconsiderable 12% (Gladwell, 2009).

Figure 1.3 Two pourable packages. One (on the right) is easier for the right-hander, the other (on the left) is better designed for the left-handed consumer. Note, though, that 90% of consumers are right-handed.

1.8 Auditory Packaging Design

While rarely given the consideration it most certainly deserves, any sounds made by the packaging when a product is picked up off the shelf, or when opened by the consumer, can play an important role in the consumer's overall multisensory product experience. Auditory cues can, for instance, be used to capture the attention of the shopper or consumer (a little too effectively in the case of the ill-fated biodegradable packaging of Sun Chips, which came in at around 100 dB when gently agitated in the shopper's hands; see Horovitz, 2010; Vranica, 2010a,b). The sounds of opening (eg, of a beverage container; Spence and Wang, 2015), or of use (think only of the sound of the aerosol spray; see Spence and Zampini, 2007) can be used to create a signature sound, one that is different from those of the opposition (such as so successfully done by the Snapple "pop"; see Byron, 2012). The sound of the product/packaging can also be designed to provide a functional benefit in terms of the consumer's overall product experience.

Thinking back to Pavlov's dogs, the distinctive sound of opening of a container can presumably also be used to set expectations in the mind of the consumer (Spence et al., 2011). Who knows, such sounds might even be capable of inducing salivation. Relevant here, a few years back, we were able to demonstrate that consumers rated potato chips as about 5% more crunchy if eaten while listening to the noisy sound of a rattling packet of crisps (Spence et al., 2011). Ideally, I would argue that packaging designers should be looking to create those packaging sounds that are both functional and distinctive (Spence, 2014). And when one considers how much money is spent on the visual aspects of branding, it is striking how so many product packages sound indistinguishable on opening (Spence and Wang, 2015); a lost marketing opportunity if ever there was one (see Byron, 2012; Spence, 2014).

1.9 Olfactory Packaging Design

Olfactory packaging design is also an intriguing area within multisensory packaging research (eg, Anon, 2010a,b; Ellison and White, 2000; Neff, 2000; Trivedi, 2006), with a growing number of companies now thinking about how best to incorporate olfactory/aroma cues into their packaging (Spence, 2015). For a number of companies, it may involve impregnating the glue with a scent-encapsulated component, so that when the consumer opens the packaging, they are hit by an aroma (from the packaging) that they will hopefully attribute to the great smell of the food or beverage product inside (see Anon, 2010a,b; Bouckley, 2013; Morran, 2013). Such an approach can be particularly beneficial for those frozen products (eg, imagine a chocolate-covered ice cream) where the low temperature suppresses any release of fragrance from the product itself (see Spence and Piqueras-Fiszman, 2012). A number of the sports waters manufacturers have already been exploring the possibility of scenting the drinking cap/spout.

One has to see it as a lost opportunity that the packaging of chocolate and tea rarely lets the consumer get a whiff of the contents prior to purchase. Here, it is interesting to contrast how effectively the coffee companies play with the scent of their product by means of incorporation of scent valves in the front of their packs. P&G, among others, has realized just how important the scent of the packaging can be to enhancing the likelihood of purchase.

1.10 Tasty Packaging

There is certainly interest in the design of edible packaging WikiPearls as one solution for yogurts, ice cream, etc. (http://www.wikipearl.com/). There is, though, always going to be a concern that consumers do not wish to eat the packaging that may have been in who knows whose hands. Previously, there have conversely been examples where the taint introduced by the packaging has become an integral part of the taste/flavor experience for the consumer, as was once apparently the case for tinned tomatoes (see Rosenbaum, 1979).

1.11 Individual/Cultural Differences in Multisensory Packaging Design

Thus far in this review, the assumption has been that all of one's potential consumers can be treated as a homogenous group. However, as the marketer knows only too well, there are important individual differences in terms of the customer base for different products. Some innovative marketers who are aware of, say, the differences in preferences between males and females, have been able to positively influence sales simply by tailoring their packaging designs to the preferences of the gender of those making the relevant purchasing decision (eg, see Cheskin, 1957). Perhaps even more important than any gender differences, though, are any cross-cultural differences in the meaning of multisensory packaging cues. While to date the focus for the limited research in this area has very much been on cultural differences in the meaning of color in packaging, one could certainly wonder whether relevant cross-cultural differences might also exist when it comes to the meaning, or influence, of product shape and/or texture (see Bremner et al., 2013 for intriguing preliminary evidence in this regard). (With regard to coloring, see Velasco et al., 2014b; Wan et al., 2014 for a couple of representative examples, and http://www.doehler.com/en/lp/multi-sensory-design-for-colours.html?utm_campaign=multi-sensory-design-for-colours&utm_medium=text-ad&utm_source=beveragedaily_website for a recent commercial example—it has for example been suggested that Cadbury's failure to break into the Japanese market could at least in part be traced back to the fact that the signature purple color of their Dairy Milk bar is associated with mourning in that part of the world.) It is also important to bear in mind here any age-related changes in the meaning and appeal of packaging color (eg, see Golletly and Guichard, 2011). Once again, this promises to be a rich area for future research.

1.12 Conclusions

The visual attributes of the packaging are perhaps the single most important sensory cue determining the success or failure of a product on the supermarket shelf. Among the various visual cues that are discernible by the consumer, color is probably the single most important attribute; it is certainly the most thoroughly studied. That said, the last few years have seen growing interest in the opportunities associated with the introduction of innovatively (or just differently) shaped packaging, a distinctive texture or finish, and/or the introduction of olfactory packaging. While the shape and feel of the packaging can be appreciated by consumer hands, as I have argued here, these features are likely to influence the consumer primarily on a visual basis. When it comes to touch and haptics, weight is perhaps the most dominant attribute (and one which cannot readily, or at least reliably, be discerned visually). Given that heavier packaging will normally equate to increased transportation costs, it remains an open question as to whether the benefits in terms of the consumer's enhanced multisensory product experience outweighs the increased shipping costs. In the case of wine, the answer would certainly seem to be in the affirmative (even though these wines can end up being shipped half way across the globe; see Piqueras-Fiszman and Spence, 2012b). In fact, in many of the cases where innovations in packaging design are being considered, there are going to be (often not insignificant) cost implications, and thus a complex calculation by the powers that be as to whether that cost is worth it in terms of increased sales/enhanced consumer product experience.

One other important question relates to how long-lasting the effects of packaging are on the consumer's multisensory product experience. Here it is certainly worth bearing in mind that the majority of studies reviewed here have been conducted over the short term. It will therefore be important for future research to look at the long-term effects in this area. Furthermore, as highlighted earlier, it is important to distinguish between the role of the packaging at the first moment of truth (Louw and Kimber, 2011), when the consumer hopefully sees and is possibly inclined to pick up the product from the shelf, and the subsequent occasion when that product is consumed, very often direct from the packaging (see Wansink, 1996).

Moving forward, excitement is starting to grow around the possibility of reinventing packaging from the bottom up based on the crossmodal correspondences that are shared by the majority of one's consumer base, likely tested and evaluated online. One example of this approach was recently outlined by Velasco et al. (2014a). The incorporation of crossmodal correspondences into the design of product packaging is an area that has grown substantially in recent years (see Schifferstein and Howell, 2015) and shows no signs of abating.

There is also growing interest from producers in considering the modification, if not the total redesign, of their product packaging so that it will stand out most effectively for those consumers who are starting to do more of their shopping online (see Spence and Gallace, 2011) or at the virtual supermarket (Moore, 2011). Under such conditions, the product packaging will likely take up a much smaller area of the customer's visual field when seen on the screen than would be the case in the supermarket aisle. This change in shopping behavior means that forward-thinking companies are increasingly trying to optimize their visual design for the screen as much as for the shelf. Here, figuring out how to bias the consumer's visual search toward the relevant section of the screen obviously becomes much more important (see Knöferle et al., submitted).

It would be remiss of me to end this piece without mentioning the growing interest in packaging that incorporates printed electronics and which is capable of capturing the consumer's attention by beeping, flashing, etc. (see Bouckley, 2014; Reynolds, 2013). There is also growing interest in functional packaging, such as the new packaging whose color actually changes to indicate when the packaging's contents have gone off (Anon 2011).

More generally, there are still important questions concerning how/whether changes in multisensory packaging design will convert into increased willingness to pay by consumers (eg, Rebollar et al., 2012; Velasco et al., 2015b), not to mention growing concern over the life span of packaging, not to mention its transportation and disposal (Lindenberg, 2012).

Finally, it is important to remember here that packaging is still just one element of the total product proposition. That is, there is also branding, labeling, etc., and it is going to be the complex interplay of all these factors that will eventually help explain the long-term success or failure of a product in the marketplace (eg, Deliza and MacFie, 2001; Mueller and Szolnoki, 2010; Nancarrow et al., 1998; Rigaux-Bricmont, 1982; Underwood and Klein, 2002).

References

Alexander, K.R., Shansky, M.S., 1976. Influence of hue, value, and chroma on the perceived heaviness of colors. Perception & Psychophysics 19, 72–74.

Anon., 1999. Touch looms large as a sense that drives sales. BrandPackaging 3 (3), 39–41.

Anon., 13 November, 2006. Retailers Promise Action on Waste. BBC News.

Anon., 21 October, 2010a. The Dutch Touch. Downloaded from http://175proof.com/triedan-dtested/the-dutch-touch/ on 19.07.15.

Anon., 2010b. Maximum appeal. Active and Intelligent Packaging World 9 (3), 4–8.

Anon., 2011. Colour-changing Food Packaging 'Could End Food Poisoning' by Showing when Fresh Produce Has Gone off. Daily Mail. Online, 22nd April. Downloaded from http://www.dailymail.co.uk/sciencetech/article-1379447/Food-poisoning-Colour-changing-packaging-shows-produce-goes-bad.html on 28.07.15.

Anon., 2013. Sure This Is the Real Thing? GREEN Coke Launched in Argentina with Natural Sweetener and Fully Recyclable Bottle. Daily Mail. Online, 22nd July. Downloaded from http://www.dailymail.co.uk/news/article-2372792/Sure-real-thing-GREEN-Coke-launched-Argentina-natural-sweetener-fully-recyclable-bottle.html on 04.01.14.

Arboleda, A.M., Arce-Lopera, C., 2015. Quantitative analysis of product categorization in soft drinks using bottle silhouettes. Food Quality and Preference 45, 1–10.

Ares, G., Deliza, R., 2010. Studying the influence of package shape and colour on consumer expectations of milk desserts using word association and conjoint analysis. Food Quality and Preference 21, 930–937.

Associated Press, 2013. Evian Revamps 'Old and Dated' Bottle after Brand Falls Behind in the Designer Water Market (But Will Anyone Spot the Difference?). Daily Mail. Online, 22nd May. Downloaded from http://www.dailymail.co.uk/news/article-2329202/Evian-revamps-old-dated-bottle-brand-falls-designer-water-market.html on 05.01.14.

Barnett, A., Spence, C. When changing the label (of a bottled beer) modifies the taste. Beverages (submitted).

Barthel, D., 1989. Modernism and marketing: the chocolate box revisited. Theory, Culture and Society 6, 429–438.

Basso, F., Robert-Demontrond, P., Hayek, M., Anton, J., Nazaian, B., Roth, M., et al., 2014. Why people drink shampoo? food imitating products are fooling brains and endangering consumers for marketing purposes. PLoS One 9 (9), e100368.

Becker, L., Van Rompay, T.J.L., Schifferstein, H.N.J., Galetzka, M., 2011. Tough package, strong taste: the influence of packaging design on taste impressions and product evaluations. Food Quality and Preference 22, 17–23.

Bertrand, K., 2002. Wake up Your Product Category with 'Shapely' Packaging. Brand Packaging (January/February).

Biggs, L., Juravle, G., Spence, C. Haptic exploration of plateware alters the perceived texture and taste of food. Food Quality and Preference (submitted).

Bititsios, S., 2012. Looks Good, Tastes Good. Downloaded from http://www.research-live.com/features/looks-good-tastes-good/4007807.article?goback=%2Enmp_*1_*1_*1_*1_*1_*1_*1_*1_*1%2Egde_136319_member_145433030 on 07.11.12.

Blythe, J., 2001. Essentials of Marketing, second ed. Prentice Hall, London, UK.

Bouckley, B., 2013. PepsiCo Seeks US Patent to Encapsulate Beverage Aromas within Packaging. Beverage Daily. 10th September. Downloaded from http://www.beveragedaily.com/Processing-Packaging/PepsiCo-seeks-US-patent-to-encapsulate-beverage-aromas-within-packaging?nocount on 24.07.15.

Bouckley, B., 2014. Dystopian Drinks? Beverages that Beep, Bleep, Yell and Waft Scents at Shoppers. Beverage Daily. 17th April. Downloaded from http://www.beveragedaily.com/Processing-Packaging/Dystopian-drinks-Beverages-that-beep-bleep-yell-and-waft-scents-at-shoppers on 07.08.14.

Bremner, A., Caparos, S., Davidoff, J., de Fockert, J., Linnell, K., Spence, C., 2013. Bouba and Kiki in Namibia? A remote culture make similar shape-sound matches, but different shape-taste matches to Westerners. Cognition 126, 165–172.

Byron, E., 2012. The Search for Sweet Sounds that Sell: Household Products' Clicks and Hums Are No Accident; Light Piano Music when the Dishwasher Is Done? The Wall Street Journal. 23rd October. Downloaded from http://online.wsj.com/article/SB10001424052970203406404578074671598804116.html?mod=googlenews_wsj#articleTabs%3Darticle on 03.09.15.

Calder, B.J., 1983. Packaging Remains an Underdeveloped Element in Pushing Consumers' Buttons. Oct 14. Marketing News, p. 3.

Chen, X., Barnes, C.J., Childs, T.H.C., Henson, B., Shao, F., 2009. Materials' tactile testing and characterisation for consumer products' affective packaging design. Materials & Design 30, 4299–4310.

Cheskin, L., 1957. How to Predict What People Will Buy. Liveright, New York, NY.

Clement, J., 2007. Visual influence on in-store buying decisions: an eye-track experiment on the visual influence of packaging design. Journal of Marketing Management 23, 917–928.

Cooper, G., 1996. Pepsi Turns Air Blue as Color Wars Reach for the Sky. The Independent. 2nd April. Downloaded from http://www.independent.co.uk/news/pepsi-turns-air-blue-as-cola-wars-reach-for-sky-1302822.html on 03.01.14.

Danger, E.P., 1987. Selecting Colour for Packaging. Gower Technical Press, Aldershot, Hants.

Davis, T., 1987. Taste tests: are the blind leading the blind? Beverage World 3 (April), 42–44 50, 85.

Day, K., 1985. Packaging Emerges as a Key Selling Tool from Cigarettes to Candy, Designers Prove that Looks Rival Content. Los Angeles Times. 17th March. Downloaded from http://articles.latimes.com/1985-03-17/business/fi-35588_1_consumer on 21.04.11.

Deliza, R., MacFie, H., 2001. Product packaging and branding. In: Frewer, L.J., Risvik, E., Schifferstein, H.N.J. (Eds.), Food, People and Society: A European Perspective of Consumers' Food Choices. Springer, Berlin, pp. 55–72.

Desanghere, L., Marotta, J.J., 2011. "Graspability" of objects affects gaze patterns during perception and action tasks. Experimental Brain Research 212, 177–187.

Dichter, E., July 1971. The Strategy of Selling with Packaging. Package Engineering Magazine. 16a–16c.

Durgee, J.F., 2003. Visual rhetoric in new product design. Advances in Consumer Research 30, 367–372.

Durgee, J.F., O'Connor, G.C., 1996. Perceiving what package designs express: a multisensory exploratory study using creative writing measurement techniques. In: Gelinas, A. (Ed.), Creative Applications: Sensory Techniques Used in Conducting Packaging Research. ASTM Publications, PA, pp. 48–61.

Elder, R.S., Krishna, A., 2012. The "visual depiction effect" in advertising: facilitating embodied mental simulation through product orientation. Journal of Consumer Research 38, 988–1003.

Ellison, S., White, E., 24 November, 2000. 'Sensory' Marketers Say the Way to Reach Shoppers Is the Nose. Wall Street Journal.

Esterl, M., 2011. A Frosty Reception for Coca-cola's White Christmas Cans. The Wall Street Journal. 1st December. Downloaded from http://online.wsj.com/article/SB100014240529 702040120045770705211375302.html on 28.10.12.

Finch, J., Smithers, R., 2006. Too Much Packaging? Dump it at Checkout, Urges Minister. The Guardian. 14th November. Downloaded from http://www.theguardian.com/business/2006/nov/14/supermarkets.ethicalliving/print on 04.01.13.

Gallace, A., Spence, C., 2014. In Touch with the Future: The Sense of Touch from Cognitive Neuroscience to Virtual Reality. Oxford University Press, Oxford, UK.

Garber Jr., L.L., Hyatt, E.M., 2003. Color as a tool for visual perception. In: Scott, L.M., Batra, R. (Eds.), Visual Persuasion: A Consumer Response Perspective. Lawrence Erlbaum, Hillsdale, NJ, pp. 313–336.

Garber Jr., L.L., Hyatt, E.M., Starr Jr., R.G., 2001. Placing food color experimentation into a valid consumer context. Journal of Food Products Marketing 7 (3), 3–24.

Gatti, E., Spence, C., Bordegoni, M., 2014. Investigating the influence of colour, weight, & fragrance intensity on the perception of liquid bath soap. Food Quality and Preference 31, 56–64.

Gimba, J.G., 1998. Color in marketing: shades of meaning. Marketing News 32 (6), 16.

Gladwell, M., 2009. What the Dog Saw and Other Conundrums. Little, Brown, & Company, USA.

Gollety, M., Guichard, N., 2011. The dilemma of flavor and color in the choice of packaging by children. Young Consumers: Insight and Ideas for Responsible Marketers 12 (1), 82–90.

Hine, T., 1995. The Total Package: The Secret History and Hidden Meanings of Boxes, Bottles, Cans, and Other Persuasive Containers. Little, Brown, and Company, New York, NY.

Horovitz, B., 2010. Frito-lay Sends Noisy, 'green' SunChips Bag to the Dump. USA Today. 10th May. Downloaded from http://www.usatoday.com/money/industries/food/2010-10-05-sunchips05_ST_N.htm.

Hruby, W.J., Sorensen, J., 1999. In P-O-P, pictures worth a thousand purchases. Marketing News 33 (24), 21–22.

Jacobs, L., Keown, C., Worthley, R., Ghymn, K.I., 1991. Cross-cultural colour comparisons: global marketers beware! International Marketing Review 8 (3), 21–31.

Johnson, A., 2007. Tactile Branding Leads Us by Our Fingertips. CTV News, Shows and Sports - Canadian Television. http://www.ctv.ca/servlet/ArticleNews/print/CTVNews/20070803/tactile_branding_070803/20070804/?hub=MSNHome&subhub=PrintStory.

Juravle, G., Velasco, C., Salgado-Montejo, A., Spence, C., 2015. The hand grasps the centre, while the eyes saccade to the top of novel objects. Frontiers in Psychology: Perception Science 6, 633.

Kampfer, K., Leischnig, A., Ivens, B.S., Spence, C. Touch-taste-transference: assessing the effect of the weight of product packaging on flavor perception and taste evaluation. International Journal of Marketing Research (submitted).

Klimchuk, M.R., Krasovec, S.A., 2013. Packaging Design: Successful Product Branding from Concept to Shelf. John Wiley & Sons, Hoboken, NJ.

Knöferle, K., Knöferle, P., Velasco, C., Spence, C. Semantically related sounds speed up visual search for products. Journal of Experimental Psychology: Applied (submitted).

Krishna, A., Morrin, M., 2008. Does touch affect taste? The perceptual transfer of product container haptic cues. Journal of Consumer Research 34, 807–818.

Labbe, D., Pineau, N., Martin, N., 2013. Food expected naturalness: impact of visual, tactile and auditory packaging material properties and role of perceptual interactions. Food Quality and Preference 27, 170–178.

Lannon, J., 1986. How people choose food: the role of advertising and packaging. In: Ritson, C., Gofton, L., McKensie, J. (Eds.), The Food Consumer. John Wiley & Sons, London, UK, pp. 241–256.

Lindenberg, L., 2012. Focusing on packaging: the unilever sustainable living plan. New Food Magazine 15 (3), 28–31.

Lindstrom, M., 2005. Brand Sense: How to Build Brands through Touch, Taste, Smell, Sight and Sound. Kogan Page, London, UK.

Louw, A., Kimber, M., 2011. The Power of Packaging. Downloaded from http://www.tnsglobal.com/_assets/files/The_power_of_packaging.pdf on 06.02.11.

Lowenthal, A.M., 1981. Design research for bilingual packaging. In: Stern, W. (Ed.), Handbook of Package Design Research. Wiley Interscience, New York, NY, pp. 399–402.

Lunt, S.G., 1981. Using focus groups in packaging research. In: Stern, W. (Ed.), Handbook of Package Design Research. Wiley Interscience, New York, NY, pp. 112–124.

Lynn, B., 1981. Color research in package design. In: Stern, W. (Ed.), Handbook of Package Design Research. Wiley Interscience, New York, NY, pp. 191–197.

Madden, T.J., Hewett, K., Roth, M.S., 2000. Managing images in different cultures: a crossnational study of color meanings and preferences. Journal of International Marketing 8 (4), 90–107.

Maison, D., Greenwald, A.G., Bruin, R., 2004. Predictive validity of the implicit association test in studies of brands, consumer attitudes, and behavior. Journal of Consumer Psychology 14, 405–415.

Marshall, D., Stuart, M., Bell, R., 2006. Examining the relationship between product package colour and product selection in preschoolers. Food Quality and Preference 17, 615–621.

Masten, L.D., 1988. Packaging's proper role is to sell the product. Marketing News 22 (2), 16.

McCabe, D., Castel, A., 2008. Seeing is believing: the effect of brain images on judgments of scientific reasoning. Cognition 107, 343–352.

McDaniel, C., Baker, R.C., 1977. Convenience food packaging and the perception of product quality: what does "hard-to-open" mean to consumers? Journal of Marketing 41 (4), 57–58.

Meyers, H.M., 1981. Determining communication objectives for package design. In: Stern, W. (Ed.), Handbook of Package Design Research. Wiley Interscience, New York, NY, pp. 22–38.

Michael, R.B., Newman, E.J., Vuorre, M., Cumming, G., Garry, M., 2013. On the (non)persuasive power of a brain image. Psychonomic Bulletin & Review 20, 720–725.

Miller, C., 1994. The shape of things: beverages sport new packaging to stand out from the crowd. Marketing News 28 (17), 1–2.

Mohan, A.M., 2013. The Sentient Side of Packaging Design. Packaging World. 1st February. Downloaded from http://www.packworld.com/package-design/structural/sentient-side-package-design on 10.01.14.

Moore, M., 6 August, 2011. Snap Purchases: The Virtual Supermarket for Busy Commuters. The Daily Telegraph, p. 17.

Morran, C., 2013. PepsiCo Thinks its Drinks Aren't Smelly Enough, Wants to Add Scent Capsules. Consumerist. 17th September. Downloaded from http://consumerist.com/2013/09/17/pepsico-thinks-its-drinks-arent-smelly-enough-wants-to-add-scent-capsules/. on 24.07.15.

Moskowitz, H., Reisner, M., Lawlor, J.B., Deliza, R., 2009. Packaging Research in Food Product Design and Development. Wiley-Blackwell, Oxford, UK.

Mueller, S., Szolnoki, G., 2010. The relative influence of packaging, labelling, branding and sensory attributes on liking and purchase intent: consumers differ in their responsiveness. Food Quality and Preference 21, 774–783.

Nancarrow, C., Wright, L.T., Brace, I., 1998. Gaining competitive advantage from packaging and labeling in marketing communications. British Food Journal 100, 110–118.

Neff, J., 2000. Product Scents Hide Absence of True Innovation. Advertising Age. 21st, 22nd February. Downloaded from http://adage.com/article/news/product-scents-hide-absence-true-innovation/59353/ on 28.11.12.

Ngo, M.K., Piqueras-Fiszman, B., Spence, C., 2012. On the colour and shape of still and sparkling water: implications for product packaging. Food Quality and Preference 24, 260–268.

Nickels, W.G., Jolson, M.A., 1976. Packaging – the fifth "p" in the marketing mix? Advanced Management Journal 41 (1), 13–21.

Nikolaidou, I., 2011. Communicating naturalness through packaging design. In: Desmet, P.M.A., Schifferstein, H.N.J. (Eds.), From Floating Wheelchairs to Mobile Car Parks. Eleven International, The Hague, pp. 74–79.

Oullier, O., 2012. Clear up this fuzzy thinking on brain scans. Nature 483, 7.

Overbeeke, C.J., Peters, M.E., 1991. The taste of desserts' packages. Perceptual and Motor Skills 73, 575–580.

Paine, F.A., Paine, H.Y., 1992. A Handbook of Food Packaging. Springer, Berlin, Germany.

Parise, C.V., Spence, C., 2012. Assessing the associations between brand packaging and brand attributes using an indirect performance measure. Food Quality and Preference 24, 17–23.

Pilditch, J., 1973. The Silent Salesman: How to Develop Packaging That Sells. Business Books, London, UK.

Pinson, C., 1986. An implicit product theory approach to consumers' inferential judgments about products. International Journal of Research in Marketing 3, 19–38.

Piqueras-Fiszman, B., Spence, C., 2011. Crossmodal correspondences in product packaging: assessing color-flavor correspondences for potato chips (crisps). Appetite 57, 753–757.

Piqueras-Fiszman, B., Spence, C., 2012a. The influence of the feel of product packaging on the perception of the oral-somatosensory texture of food. Food Quality and Preference 26, 67–73.

Piqueras-Fiszman, B., Spence, C., 2012b. The weight of the bottle as a possible extrinsic cue with which to estimate the price (and quality) of the wine? Observed correlations. Food Quality and Preference 25, 41–45.

Piqueras-Fiszman, B., Spence, C., 2012c. The weight of the container influences expected satiety, perceived density, and subsequent expected fullness. Appetite 58, 559–562.

Piqueras-Fiszman, B., Spence, C., 2012d. Sensory incongruity in the food and beverage sector: art, science, and commercialization. Petits Propos Culinaires 95, 74–118.

Piqueras-Fizman, B., Spence, C., 2015. Sensory expectations based on product-extrinsic food cues: an interdisciplinary review of the empirical evidence and theoretical accounts. Food Quality and Preference 40, 165–179.

Piqueras-Fiszman, B., Velasco, C., Salgado-Montejo, A., Spence, C., 2013. Combined eye tracking and word association analysis to evaluate the impact of changing the multisensory attributes of food packaging. Food Quality and Preference 28, 328–338.

Plasschaert, J., 1995. The meaning of colour on packaging – a methodology for qualitative research using semiotic principles and computer image manipulation. In: Decision Making and Research in Action. 48th ESOMAR Marketing Research Congress, pp. 217–232 (Amsterdam, The Netherlands).

Pradeep, A.K., 2010. The Buying Brain: Secrets of Selling to the Subconscious Mind. Wiley, Hoboken, NJ.

Prince, G.W., May 31, 1994. The contour: a packaging vision seen through coke-bottle lenses. Beverage World 113, 1–6 (Periscope Edition).

Prone, M., 1993. Package Design Has Stronger ROI Potential than Many Believe. Marketing News, p. 13.

Raine, T., 2007. Multisensory Appeal. May. Packaging News, pp. 36–37.

Rebollar, R., Lidón, I., Serrano, A., Martín, J., Fernández, M.J., 2012. Influence of chewing gum packaging design on consumer expectation and willingness to buy. An analysis of functional, sensory and experience attributes. Food Quality and Preference 24, 162–170.

Reynolds, P., 2013. Electroluminescent Packaging. Packaging World. 3rd July. Downloaded from http://www.packworld.com/package-design/package-manufacturing-advances/electroluminescent-packaging. on 10/01/2014.

Rigaux-Bricmont, B., 1982. Influences in brand name and packaging on perceived quality. Advances in Consumer Research 9, 472–477.

Rigby, R., 2010/2011. Perfume Industry: Boxing Clever. Raconteur, p. 14.

Rosenbaum, R., 1979. Today the strawberry, tomorrow…. In: Klein, N. (Ed.), Culture, Curers and Contagion. Chandler & Sharp, Novato, CA, pp. 80–93.

Roullet, B., Droulers, O., 2005. Pharmaceutical packaging color and drug expectancy. Advances in Consumer Research 32, 164–171.

Sacharow, S., 1970. Selling a package through the use of color. Color Engineering 9, 25–27.

Sacharow, S., 1982. The Package as a Marketing Tool. Chilton Books, Radnor, PA.

Salgado-Montejo, A., Velasco, C., Maya, C., Spence, C., 2014. La ciencia del color y cómo se puede aplicar a envases y etiquetas (The science of colour and how it can be applied to packaging and labels). El Empaque + Conversión 20 (2), 46–48.

Schifferstein, H.N.J., Fenko, A., Desmet, P.M.A., Labbe, D., Martin, N., 2013. Influence of packaging design on the dynamics of multisensory and emotional food experience. Food Quality and Preference 27, 18–25.

Schifferstein, H.N.J., Howell, B.F., 2015. Using color-odor correspondences for fragrance packaging design. Food Quality and Preference 46, 17–25.

Schlossberg, H., August 6, 1990. Effective Packaging 'talks' to Consumers. Marketing News, pp. 6–7.

Seher, T., Arshad, M., Ellahi, S., Shahid, M., 2012. Impact of colors on advertisement and packaging on buying behavior. Management Science Letters 2, 2085–2096.

Simms, C., Trott, P., 2010. Packaging development: a conceptual framework for identifying new product opportunities. Marketing Theory 10, 397–415.

Smets, G.J.F., Overbeeke, C.J., 1995. Expressing tastes in packages. Design Studies 16, 349–365.

Spence, C., 2009. Measuring the impossible. In: MINET Conference: Measurement, Sensation and Cognition. National Physical Laboratories, Teddington, UK, pp. 53–61.

Spence, C., 2011. Crossmodal correspondences: a tutorial review. Attention, Perception, & Psychophysics 73, 971–995.

Spence, C., 2012. Managing sensory expectations concerning products and brands: capitalizing on the potential of sound and shape symbolism. Journal of Consumer Psychology 22, 37–54.

Spence, C., 2014. Multisensory advertising & design. In: Flath, B., Klein, E. (Eds.), Advertising and Design. Interdisciplinary Perspectives on a Cultural Field. Verlag, Bielefeld, pp. 15–27.

Spence, C., 2015. Leading the consumer by the nose: On the commercialization of olfactory-design for the food and beverage sector. Flavour 4, 31.

Spence, C., Gallace, A., 2011. Multisensory design: reaching out to touch the consumer. Psychology & Marketing 28, 267–308.

Spence, C., Piqueras-Fiszman, B., 2011. Multisensory design: weight and multisensory product perception. In: Hollington, G. (Ed.), Proceedings of RightWeight, vol. 2. Materials KTN, London, UK, pp. 8–18.

Spence, C., Piqueras-Fiszman, B., 2012. The multisensory packaging of beverages. In: Kontominas, M.G. (Ed.), Food Packaging: Procedures, Management and Trends. Nova Publishers, Hauppauge NY, pp. 187–233.

Spence, C., Shankar, M.U., Blumenthal, H., 2011. 'Sound bites': auditory contributions to the perception and consumption of food and drink. In: Bacci, F., Melcher, D. (Eds.), Art and the Senses. Oxford University Press, Oxford, UK, pp. 207–238.

Spence, C., Wang, Q., 2015. Sonic expectations: On the sounds of opening and pouring. Flavour 4, 35.

Spence, C., Zampini, M., 2007. Affective design: modulating the pleasantness and forcefulness of aerosol sprays by manipulating aerosol spraying sounds. CoDesign 3 (Suppl. 1), 109–123.

Spinney, L., 18 September, 2013. Selling Sensation: The New Marketing Territory. New Scientist. 2934.

Stern, W. (Ed.), 1981. Handbook of Package Design Research. Wiley Interscience, New York, NY.

Stoll, M., Baecke, S., Kenning, P., 2008. What they see is what they get? An fMRI-study on neural correlates of attractive packaging. Journal of Consumer Behaviour 7, 342–359.

Stuckey, B., 2012. Taste What You're Missing: The Passionate Eater's Guide to Why Good Food Tastes Good. Free Press, London, UK.

Trivedi, B., 2006. Recruiting Smell for the Hard Sell. New Scientist 2582, pp. 36–39.

Tu, Y., Yang, Z., Ma, C., 2015. Touching tastes: the haptic perception transfer of liquid food packaging materials. Food Quality and Preference 39, 124–130.

Tutssel, G., 2001. But You Can Judge a Brand by Its Colour. Nov. Brand Strategy, pp. 8–9.

Underwood, R.L., 1993. Packaging as an extrinsic product attribute: an examination of package utility and its effect on total product utility in a consumer purchase situation. In: Varadarajan, R., Jaworski, B. (Eds.), Marketing Theory and Applications, vol. 4. American Marketing Association, Chicago, IL, pp. 212–217.

Underwood, R.L., Ozanne, J., 1998. Is your package an effective communicator? A normative framework for increasing the communicative competence of packaging. Journal of Marketing Communication 4, 207–220.

Underwood, S., Klein, N., 2002. Packaging as brand communication: effects of product pictures on consumer responses to the package and brand. Journal of Marketing Theory and Practice 10 (4), 58–68.

Usborne, S., 22 November, 2012. Why We Shrink-wrap the Cucumber. The Independent, pp. 42–43.

Vartan, C.G., Rosenfeld, J., August 1987. Winning the Supermarket War: Packaging as a Weapon. Marketing Communications, p. 33.

Velasco, C., Salgado-Montejo, A., Marmolejo-Ramos, F., Spence, C., 2014a. Predictive packaging design: tasting shapes, typographies, names, and sounds. Food Quality and Preference 34, 88–95.

Velasco, C., Wan, X., Salgado-Montejo, A., Woods, A., Andrés Oñate, G., Mu, B., Spence, C., 2014b. The context of colour-flavour associations in crisps packaging: a cross-cultural study comparing Chinese, Colombian, and British consumers. Food Quality and Preference 38, 49–57.

Velasco, C., Woods, A.T., Hyndman, S., Spence, C., 2015a. The taste of typeface. i-Perception 6 (4), 1–10.

Velasco, C., Woods, A.T., Spence, C., 2015b. Evaluating the orientation of design elements in product packaging using an online orientation task. Food Quality and Preference 46, 151–159.

Vranica, S., 10 August, 2010a. Snack Attack: Chip Eaters Make Noise about a Crunchy Bag Green Initiative Has Unintended Fallout: A Snack as Loud as 'the Cockpit of My Jet'. Downloaded from Wall Street Journal. http://online.wsj.com/news/articles/SB100014240 5274870396000457542715010329390 6 on 24.07.14.

Vranica, S., 6 October, 2010b. Sun Chips Bag to Lose Its Crunch. The Wall Street Journal. Downloaded from http://online.wsj.com/article/SB10001424052748703843804575534182403878708.html.

Wan, X., Woods, A.T., van den Bosch, J., Mckenzie, K.J., Velasco, C., Spence, C., 2014. Cross-cultural differences in crossmodal correspondences between tastes and visual features. Frontiers in Psychology: Cognition 5, 1365.

Wan, X., Woods, A.T., Velasco, C., Salgado-Montejo, A., Spence, C., 2015. Assessing the expectations associated with pharmaceutical pill colour and shape. Food Quality and Preference 45, 171–182.

Wansink, B., 1996. Can package size accelerate usage volume? Journal of Marketing 60, 1–15.

Weinstein, S., 1981. Brain wave analysis: the beginning and future of package design research. In: Stern, W. (Ed.), Handbook of Package Design Research. Wiley Interscience, New York, NY, pp. 492–504.

Weisberg, D.S., Keil, F.C., Goodstein, J., Rawson, E., Gray, J.R., 2008. The seductive allure of neuroscience explanation. Journal of Cognitive Neuroscience 20, 470–477.

Westerman, S.J., Sutherland, E.J., Gardner, P.H., Baig, N., Critchley, C., Hickey, C., Zervos, Z., 2013. The design of consumer packaging: effects of manipulations of shape, orientation, and alignment of graphical forms on consumers' assessments. Food Quality and Preference 27, 8–17.

Wheatley, J., 24–29 October, 1973. Putting Colour into Marketing. Marketing, p. 67.

Whyte, S., 12 July, 2013. Good Buys with Terrible Twists. The Sydney Morning Herald, p. 5 (News).

Woods, A.T., Velasco, C., Levitan, C.A., Wan, X., Spence, C., 2015. Conducting perception research over the internet: a tutorial review. PeerJ 3, e1058.

Zampini, M., Mawhinney, S., Spence, C., 2006. Tactile perception of the roughness of the end of a tool: what role does tool handle roughness play? Neuroscience Letters 400, 235–239.

Consumer Reactions to On-Pack Educational Messages

2

K.G. Grunert
Aarhus University, Aarhus, Denmark

2.1 The Food Label as an Information Source

Most decisions on the purchase of food products are made in the shop, and the appearance of food products on the shelf has therefore long been acknowledged as a major factor impacting purchases. This includes the shape and design of the packaging, and it includes the food label and all communicational elements on it. Food labels have always been used not only to provide information on the brand name and on the type of product, but also to advertise special desirable characteristics of the product, to trigger positive associations, and to provide information on the product, both mandatory and voluntary.

Today, the amount of information on food labels can be daunting. There are three major reasons for this. First, the quality of a food product is, before the purchase, largely unknown (Grunert, 2005). Taste, the number one quality parameter that consumers demand in food, is an experience quality that is revealed after but not before the purchase, and good taste therefore needs to be signaled by the package and label. Second, and more importantly, taste has been supplemented by a host of additional quality parameters used in the marketing of food, many of which are intangible and therefore need to be communicated (Fernqvist and Ekelund, 2014). This includes healthfulness, sustainability, and authenticity, all of which are complex, difficult to communicate parameters. They are in demand by consumers (at least some consumers), but their potential use for marketing and positioning depends on food producers' abilities to provide clear and credible communication on these parameters. Third, and related to the previous, many of these parameters also have a public policy component attached. In the wake of the obesity issue and other diet-related health problems, healthy eating has become a societal goal. If consumers are to be encouraged to make healthier choices, they need to be informed about which products are more and which are less healthful, and information on the food label is one way of doing this. Likewise, there has been a growing conviction that consumer food choices have a major impact on the sustainability of food production, resulting in a societal goal of encouraging more sustainable food choices, which again can be supported by information on the food label. In many countries, there is also a policy of encouraging organic, local, artisanal, or domestically produced food, again information that needs to be communicated on the food label.

Much information on food labels these days therefore has a dual role. It serves the purpose of positioning the food product in line with the marketing strategy of the

Integrating the Packaging and Product Experience in Food and Beverages. http://dx.doi.org/10.1016/B978-0-08-100356-5.00002-4

food producer, and it is a way of supporting public policy. Public policy can make certain pieces of information mandatory, and it can regulate those that are voluntary. Public policy can also encourage food producers to engage in voluntary information provision schemes that they adopt in support of their own corporate social responsibility policies.

This information comes in many forms. In terms of content, the major categories are nutrition labeling, health and nutrition claims, and sustainability information. In terms of format, it ranges from small print hidden away on the back of the package to ingeniously designed symbols prominently displayed on the front. Because of the duality of purposes of this information, it is not surprising that there has been some debate on its effects and how they depend on the kind of formatting. While many questions in this respect are still unanswered and there is a good deal of ongoing research, there is by now also a body of knowledge that can help us in forming opinions about what can and what cannot be realistically achieved by this kind of information on food labels.

This chapter will give an introduction to the major kinds of on-pack educational messages on food products and their likely effects. By "educational messages," we designate all kinds of information on the food label that, in addition to a selling purpose, are also intended to have an effect on consumers that in some way is regarded as socially desirable. The major categories of such messages are the ones listed above: nutrition labeling, health and nutrition claims, and sustainability information. We will, in the following, first present and discuss a model of consumer processing of on-pack information, which will give some direction with regard to which types of effects we can expect and what they are contingent on. We will then go through the three major types of messages listed above and summarize major findings with regard to their effects. After that, we will briefly address the complex issue of context effects—how the food label context in which the message is embedded can affect the effects. We will close by highlighting some new developments in package communication.

2.2 A Model of Consumer Processing of On-Pack Information

It is common to analyze the effects of market communication from a hierarchy-of-effects perspective (Barry, 1987). A hierarchy of effects is a sequence of effects, each of which is a necessary, but not sufficient prerequisite for the next effect to occur. Together, they provide a chain of effects leading up to some final focal effect. In marketing, where hierarchy of effects models have been developed, the final focal effect is the purchase of the product (brand choice). However, from a public policy perspective, we may want to look for additional effects. For example, nutrition labeling may result in a consumer choosing a healthier variety among a series of breakfast cereal products, but if the consumer then puts more sugar on the cereal than s/he otherwise would have done, the potential health effect may vanish. From a public policy perspective, it therefore makes sense not to stop with the brand choice, but to look at potential effects on the pattern of purchases and on the pattern of consumption (Hieke et al., 2015).

In the model to be presented now, we therefore have both the brand choice and the consumption pattern as final effects. We should warn, though, that considerably more is already known about brand choice effects than about effects on consumption patterns.

To understand the impact of educational messages on brand choice and consumption patterns, a broad range of theoretical concepts may be invoked. One simple way of structuring the problem area is to use a dual processing hierarchy-of-effects framework, as illustrated in Fig. 2.1 (adapted from Grunert et al., 2012a; see also Balasubramanian and Cole, 2002; Grunert and Wills, 2007; Moorman, 1990). Consumers need to be exposed to the message; otherwise, no further effects will occur. Exposure of front-of-pack (FOP) messages may be accidental, because the consumer in most cases looks at a product before selecting it (Grunert et al., 2010). However, for messages placed on the back of the package, exposure will require some effort from the consumer, because the product has to be turned around. Perception of messages on the back of the pack is therefore most likely dependent on the consumer's motivation and ability to process the message, whereas perception of FOP information also will be affected by the attention-getting properties of the message. If the message is indeed perceived, further processing can follow two paths.

Path 1 is cognitively dominated and involves conscious efforts to assign meaning to the message. The process of assigning meaning can be subdivided into *understanding* and *inferences*, although the distinction between the two is sometimes a bit blurred. Understanding refers to the manifest content of the message, whereas inferences refer to the consumer's attempt to relate the message to self-relevant consequences—for example, whether the nutrition claim involves less energy and hence the potential for staying slim, whether the health claim will lead to a lower chance of catching a cold in the winter, or whether the locally produced product will sustain local business. Both understanding and inferences depend on previous knowledge of the consumer and on previous experiences in processing this type of information. Inferences may then enter the decision-making process, where they may be used as input in a decision heuristic or traded off against other criteria like taste, family liking, convenience, and price.

Path 1 in Fig. 2.1 traces the cognitive effects of the on-pack message, and processing via this path will depend on the levels of both motivation and ability to process the

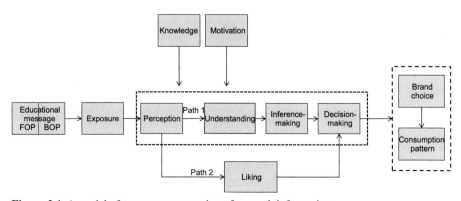

Figure 2.1 A model of consumer processing of on-pack information.

message when buying food. However, food is frequently bought, so some product categories may be low-involvement purchases for the consumer, and many brand choices may be habit-based. In such cases the educational message may just be ignored, or it may have affective effects as described by Path 2 in Fig. 2.1. Some educational messages may elicit affective responses even if they are not processed at any depth—a heart-shaped health logo may elicit feelings of warmth and affection, and the presence of green or red traffic lights on the front of the pack may elicit positive or negative emotions that impact brand choice without further cognitive processing. The literature has typically not distinguished between these paths and has tended to focus on cognitive responses; research on affective response to educational messages is scant.

As noted above, from a public policy perspective, it is often not sufficient to obtain effects on brand choice alone. The healthier or more sustainable brand choice is fine, but its effects on overall healthiness of the consumer's diet or on the overall sustainability of that consumer's consumption pattern will depend on whether the (small) positive effect on brand choice will or will not be offset by changes in all other aspects of buying, preparing, and eating food. In addition to the effect on brand choice, and taking nutritional messages as an example, the message may affect the quantity eaten within the category (eg, consumers may eat more of a product if they perceive it as low fat), across categories (eg, eating healthier during regular meals may lead consumers to think they can indulge more into snacking), and even changes in meal patterns and eating habits (eg, going out for dinner instead of eating ready meals after realizing the latter are high in fat and salt). The single-brand choice represents one of a large number of decisions that have an impact on dietary intake or sustainable consumption patterns. In addition to the totality of brand choices in the food area, it is influenced by decisions on menus, meal preparation methods, choice of recipes, eating in vs. eating out, meal patterns, and snacking habits. While many of these may not be directly affected by educational messages, indirect effects may occur that either reinforce or counteract desirable effects at the brand choice level.

Compared with explaining brand choices, the body of consumer behavior theory we can draw upon is considerably sparser when it comes to explaining consumption patterns at the aggregate level. Economic models of demand have possible substitutions between various product categories built in, and while relative prices are viewed as the primary determinant of demand in economics, effects of information also can be analyzed as part of demand models (see, eg, Mazocchi et al., 2009). Sociological approaches have also been employed to analyze changes in, for example, meal patterns in response to changes in society, and the introduction of educational messages can be viewed as an aspect of a changing societal discourse on issues like health and sustainability that also impacts what consumers buy and eat (eg, Mennel et al., 1992).

2.3   Effects of Major Types of On-Pack Messages

2.3.1   Nutrition Labeling

Nutrition labeling refers to information about the nutritional content of a food product. In many countries of the world (including the European Union and the United States), it is mandatory to provide information on energy content and the content of key nutrients (like fat, saturated fat, sugar, and salt) per 100 g and/or per portion size. This mandatory information is most often provided in small print on the back of the pack. In addition, a range of voluntary FOP labeling schemes have been introduced. Most of these are likewise key nutrient–based, giving information on energy and key nutrient contents in more prominent format on the front of the pack. This information is often supplemented by information on the guideline daily allowance (GDA), a percentage showing how much of the recommended maximum daily intake of a nutrient is contained in a portion or 100 g of the product. Another scheme, used especially in the United Kingdom, is traffic light labeling, where the key nutrient information is supplemented by a green, amber, or red color, depending on the amount of the nutrient contained in the product, with green light indicating a healthier alternative, a red light a treat that should be consumed only occasionally, and the amber in between. In addition to these key nutrient schemes, there are health symbol schemes. A health symbol is a symbol that a product is allowed to carry if it lives up to a certain nutritional profile. Health symbols may be state-run (like the Scandinavian "keyhole") or run by an independent body (like the "Choices" logo), and may be used for all or only for certain product categories. They are usually dichotomous—the product does or does not carry the symbol—but there are examples of graded symbols (eg, a product having between zero and three stars). Examples of various nutrition labeling schemes are shown in Fig. 2.2.

In the public debate about nutrition labeling, a lot of attention has been given to issues of consumer understanding, especially to the question of whether certain formats (like traffic light labeling) are easier to understand than others (like GDA labeling). Today, after about a decade of research on the topic, it is widely accepted that understanding is not the major problem (Grunert et al., 2012b). Several studies have shown that when consumers are given two or more products (from the same product category) with key nutrient information, consumers usually can find out which one is the healthier alternative, and this does not depend on whether the information is given in one or another format (Grunert et al., 2010; Hodgkins et al., 2015; Mahlam et al., 2009). Most consumers have learned that more fat, salt, and sugar means less healthy, and they can compare numbers in terms of higher or lower. Understanding can be more of an issue when consumers try to compare products from different categories, but may be alternatives with regard to a common goal (say, eating an apple or a muesli bar for a snack)—in this case, differences in portion sizes (both those used to compute the information and those actually used by the consumer) can complicate the comparison, and consumers may find some of the information at odds with their preconceived notions about the relative healthfulness of different product categories (Raats et al., 2015).

Even though consumer understanding thus does not seem to be a bottleneck, evidence on actual effects on purchase decisions is quite limited. An American study on the introduction of a graded (zero to three stars) health symbol on all products in a retail chain showed that over 2 years, the share of products in shoppers' baskets that had at least one star rose by about 2% (Sutherland et al., 2010). A study based on scanner data from a major UK retail chain, which analyzed effects of the introduction of a GDA-based key nutrient system on all its private label products, indicated that the introduction may have influenced the market share of less healthful product variants

(A)

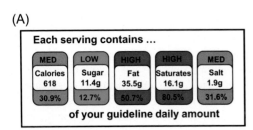

System containing energy and key
nutrient information, % guideline
daily amounts and traffic light colors

System containing energy and key
nutrient information and traffic light colors

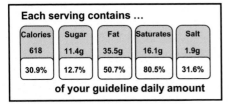

System containing energy and key
nutrient information and % guideline
daily amounts

(B)

'Keyhole': Government system for
identifying products
that are healthiest within
a product category

**BRITISH HEART
FOUNDATION**

Heart Foundation system for
identifying products that are
'heart healthy' within a
product category

Food industry system for
identifying products that are
healthiest within a
product category

Figure 2.2 Examples of nutrition labeling schemes—key nutrient information with/without guideline daily allowances and traffic light labeling (A) and health symbols (B).

but did not indicate any effect on brand choice (Boztug et al., 2015). Likewise, another study in the United Kingdom, on effects of introducing a traffic light–based key-nutrient scheme by another major retailer, showed no effects when comparing sales 4 weeks before and after introduction for selected product categories (Sacks et al., 2009). The cautious conclusion from these studies is that the effects of nutrition labeling schemes on brand choice are bound to be small.

If understanding is not the major bottleneck, why then do these schemes not have a bigger impact on brand choice? Research suggests that the major reasons are lack of attention and lack of motivation (Grunert et al., 2012b).

Attention is a prerequisite for any further processing and hence also for any effects on brand choice (Bialkova et al., 2014). Decisions in supermarkets are made within seconds, and much information on food labels therefore goes unnoticed. Research using mobile eye tracking in supermarkets indicates that the information most likely to be attended is the brand, the product name, and the pictorial stimulus on the pack, and that all other information has a relatively low probability of being attended (Groeppel-Klein, 2011), a result corroborated by another study using a combination of observation and interviewing (Grunert et al., 2010).

Attention can be increased by the visual salience of the information, which in turn is related mainly to size and the way the information contrasts with its surroundings (Graham et al., 2012). Attention is also related to the amount of clutter on the label (Bialkova et al., 2013), and it can be increased by having the information consistently always at the same place in the same format. When shoppers are under time pressure, simple information—like a health symbol—is more likely to be attended than a more complicated, key nutrient–based format.

Motivation is a bottleneck, but not because consumers are uninterested in their health. In a food context, the health motive may simply not be salient at the time of purchase, and even when it is salient it may be traded off against other relevant decision criteria, like the taste of the product or family liking (Grunert, 2005). As already noted, much food purchasing is habitual or spontaneous and not subject to a lot of conscious deliberation, which could have resulted in the health goal becoming activated. Lab-based research has consistently shown that the impact of nutrition labeling on choices increased (sometimes dramatically) when the health motive was made salient for the participants (eg, Bialkova et al., 2014).

While most consumers thus seem to be able to understand nutrition labeling messages, their impact on actual purchase is limited due to limited attention and lacking motivation. This applies to effects via the cognitive route (Path 1 in Fig. 2.1). We know little about possible effects along Path 2. One can speculate that the effects of nutrition labeling will rise slowly over time as general health awareness increases.

2.3.2 Health and Nutrition Claims

Health and nutrition claims are claims about the specific nutrition or health-related properties of the product. A nutrition claim is a statement about a specific nutrient, claiming that this nutrient is contained in the product ("contains vitamin C"), that the content of the nutrient is high or low ("low in fat"), or that the content of the nutrient is higher or lower compared with some comparative standard ("now 20% less salt"). A health claim links a particular ingredient of a food product to physiological functions of the body ("omega-3 helps keep your arteries clean…") and/or a health benefit ("…which reduces the risk of heart disease"). Nutrition and especially health claims are usually tightly regulated, with specific wordings prescribed and narrow constraints for any changes in the prescribed wording.

Health claims became the focus of attention in the context of functional foods, that is, foods which have specific health-related benefits beyond their normal nutritive function. For some time, functional foods, usually developed by enriching some carrier food with a specific functional ingredient, were regarded as the major avenue to growth for the food industry. More recently the hype has subsided, not only because of narrow legislative constraints, but also because consumer acceptance has in many cases been lukewarm.

As already noted, any health benefit of a food product needs to be communicated, and this holds especially for food with specific and special health properties, which need to be communicated by a health claim. Again, we can analyze the effects of nutrition and health claims on consumers by looking at their ability to attract attention and be understood, and the inferences consumers make from them before possibly being used in decision-making, and we can relate these effects to consumer motivation and knowledge.

The general constraints regarding attention and the ways in which the format of the information can try to get around these, discussed in the preceding section, also hold here. But as attention is related to motivation, a major difference to nutrition labeling should be noted here. Healthy eating is a motive that (to varying extent) is relevant to most consumers, even though it may not be salient and hence not behaviorally relevant in a particular choice situation. Therefore, most consumers have some latent motivation for processing nutrition information. The same is not true for health claims, which refer to a specific health effect that may or may not be perceived as relevant by a particular consumer. A health claim promising low levels of cholesterol is relevant for people having a cholesterol problem, but probably not for others. Perceived relevance is a major motivational factor with regard to health claims (Dean et al., 2012), and of course it is the perceived and not the objective relevance that is of importance, as the recent craze about gluten-free products clearly demonstrates.

Many health claims are formulated in technical and not very accessible language. Even though the EU legislation actually requires that a health claim must be understandable by the "average consumer," not too much is actually known about consumer ability to understand health claims. Consumer inferences beyond the manifest claim have received more attention in the literature, driven by a concern that consumers may draw inferences that go far beyond what the claim promises. The results have been rather mixed. US-based research has found evidence for what has come to be called the "magic bullet effect," where a product with a health claim is taken to be a general cure for all kinds of diseases (Roe et al., 1999), but research in Europe could not replicate this effect and even found an opposite effect, where the presence of a health claim on a fortified product was taken as an indicator of a less healthy product—a result probably due to the skepticism of many European consumers regarding all kinds of manipulations of food products (Lähteenmäki et al., 2010). One way or another, understanding and inferences are bigger issues with health claims than they are with nutrition labeling, but even here research shows that motivation is a bigger factor in explaining consumers' inclinations to use health claims in their shopping than is their ability to understand them (Hoefkens et al., 2015).

There are few studies investigating the effect of health claims on consumer choice in a real-world setting, and those that have suggested that consumers largely ignore health claims (Aachmann et al., 2013). Not surprisingly, lab studies involving forced exposure to health claims find more effects, with a main finding that the attractiveness of any particular health claim is subject to individual differences in the perceived relevance of the claim (Wills et al., 2012). Also investigated has been how various aspects of the formulation of the claim affect its effect—for example, complexity of the claim, positive or negative framing, with or without qualifiers—but again results are not consistent, depending on both the consumer and the particular claim (Grunert et al., 2009).

Again, most analyses of the effect of nutrition and health claims follow, explicitly or implicitly, Path 1 in Fig. 2.1. We cannot rule out that certain types of claims may elicit affective reactions that are not cognitively mediated and may have an effect on purchases. Generally speaking, though, it seems that the effect of health claims is highly contingent on whether the claim is relevant and familiar to the consumer, which in turn will differ among different consumers.

2.3.3 Sustainability Information

Compared to the healthy eating issue, which has been on the public agenda for at least half a century, the sustainable choices issue is of relatively newer origin. Sustainability-related information on food labels has been triggered by an increasing conviction that consumer choice is an important part in bringing about more sustainable food production, supplemented by industry activities aimed at using sustainable production as a parameter that can support corporate and brand image in the eyes of consumers and other stakeholders. As a result, a wealth of sustainability-related labels and logos have appeared on food products, of which the more prominent ones are various organic labels, the Fair Trade label, the Rain Forest Alliance label, various carbon index schemes, and animal welfare–related logos.

Sustainability is a broad, complex, and for many consumers diffuse concept. While the tripartite definition of sustainability as having environmental, social, and economic aspects has gained wide acclaim both in the literature and in policy development, many consumers have a more narrow view. Research conducted in several European countries indicated that consumers associate sustainability mostly with environmental considerations and with intergenerational justice, and to a lesser extent with ethical and social issues (Grunert et al., 2014). As a consequence, sustainability information on food labels mostly refers to a particular aspect of sustainability and not to sustainability in general.

Most aspects discussed above with regard to the effects of health-related messages on food labels hold here as well. This goes especially for the attention-getting properties of the messages. However, there are some important differences. While nutrition information is relatively standardized and restricted to a few basic formats, sustainability information comes in a plethora of different forms, with more than 100 different schemes available in the European Union alone (European Commission, 2012). Consumer awareness of many of these will therefore be low, and even for the more widespread ones, like the Rainforest Alliance scheme, awareness differs considerably between countries. A lack of awareness implies a lack of knowledge on what the scheme stands for, which in turn is a barrier for understanding. Some schemes appear quite self-explanatory, but it has also been shown that even well-known schemes like the Rainforest Alliance are often misinterpreted (Grunert et al., 2014). In addition, some labeling schemes are used as a basis for inferences which go far beyond what the label stands for. This is well documented for organic labels, which are widely viewed as a general quality mark, such that organic products are believed to be superior in terms of all major quality parameters including taste and healthfulness (eg, Zanoli and Naspetti, 2002). Even more narrow schemes like Fair Trade lead some consumers to believe that the product has a superior taste (de Ferran and Grunert, 2007). Once such inferences become widespread, they increase the motivational base for using such information in decision-making, but at the same time lead to an increased risk of consumer disappointment if the high expectations are not fulfilled.

The organic label stands out as it is actively being used by consumers in their decision-making, and for some consumers has even reached the status of a decision heuristic at the same level as brand or price (Thøgersen et al., 2012). The impact of most other sustainability labels on actual consumer decision-making is more doubtful. Many consumers are in principle motivated to behave sustainably, but it seems that this motive does not often become behaviorally relevant in a choice situation where other motives are more salient. Even when sustainability considerations actually enter the decision-making process, they may be traded off against other parameters like price, taste, and even health-related issues (Grunert et al., 2014).

The role of sustainability messages on food packages is bound to increase in the future. The future role of sustainability messages on food labels will depend on a host of factors, though, including the general prominence of sustainability issues in the public debate and the ability of sustainability labels to achieve the status of broad quality marks as is currently the case with organic labels.

2.4 The Role of Context

In the discussion above, we have treated health- and sustainability-related messages on food labels in isolation. However, they are of course embedded in the food label, which contains other information and pictorial stimuli which are framed by the label design. The effect of both health- and sustainability-related messages is bound to depend on the context in which they appear.

One could even turn this argument around and argue that nutrition information, health claims, and ecolabels are part of the context, whereas the core messages on a food label are something else. We noted above that, in a time-pressured shopping situation, the elements of the food label most likely to be noticed are the brand, the product name, and the picture. Research on the perception of the healthfulness of food products has shown that the choice of picture on the label has considerably more influence on the perception of healthfulness than a health or nutrition claim (and that even information about organic or local production has more impact on the perception of healthfulness; see Chrysochou and Grunert, 2014), and research on package design generally supports the notion that brand, imagery, and color have an important impact on how the product is perceived. The more that consumers follow Path 2 in Fig. 2.1 rather than the more cognitively demanding Path 1, the stronger these effects are likely to be.

2.5 New Developments in Package Communication

The food label is subject to severe constraints as a communication channel. Many food labels already appear overloaded, and the developments sketched in the beginning of this chapter point in the direction of food products becoming still more information rich in the future.

Not surprisingly then, there has been discussion about how the food label can be supplemented by other information channels. Already, much information about food products is available on the Internet, and it is common for food labels to refer consumers to websites, for example, by QR codes. Likewise, the proliferation of smartphones has already made it possible to obtain additional information at the point of purchase, and a range of apps has been developed that can help consumers in making informed decisions.

To date, none of these has revolutionized the way in which consumers shop. A study in the United States on the adoption of such a shopping app showed that of the several functions this app had, the only one that received widespread acceptance was to use it for making shopping lists (Shi and Zhang, 2014). This lack of impact of much of the new technology-based solutions for providing shopping-related information is most likely related to the fact that almost all these devices further increase instead of decrease the information load under which consumers do their shopping. A classical store is a highly information-rich environment, and additional information being supplied by handheld devices only makes the shopping task more difficult for consumers, *unless* they include a way of reducing the information overload by tailoring the information stream to the individual consumer. Information provision systems that restrict the information stream based on individual preferences, such that only information of relevance for the consumer gets through the filter, could revolutionize the way in which consumers shop, including the ways in which educational messages impact their choices.

References

Aachmann, K., Hansen, I.H., Grunert, K.G., 2013. Ernærings-og sundhedsanprisninger-forståelse og anvendelse blandt danske forbrugere. DCA-Nationalt center for fødevarer og jordbrug.

Balasubramanian, S.K., Cole, C., 2002. Consumers' search and use of nutrition information: the challenge and promise of the nutrition labeling and education act. Journal of Marketing 66 (3), 112–127.

Barry, T.E., 1987. The development of the hierarchy of effects: an historical perspective. Current Issues and Research in Advertising 10, 251–295.

Bialkova, S., Grunert, K.G., van Trijp, H., 2013. Standing out in the crowd: the effect of information clutter on consumer attention for front-of-pack nutrition labels. Food Policy 41, 65–74.

Bialkova, S., Grunert, K.G., Juhl, H.J., Wasowicz-Kirylo, G., Stysko-Kunkowska, M., van Trijp, H.C., 2014. Attention mediates the effect of nutrition label information on consumers' choice. Evidence from a choice experiment involving eye-tracking. Appetite 76, 66–75.

Boztuğ, Y., Juhl, H.J., Elshiewy, O., Jensen, M.B., 2015. Consumer response to monochrome guideline daily amount nutrition labels. Food Policy 53, 1–8.

Chrysochou, P., Grunert, K.G., 2014. Health-related ad information and health motivation effects on product evaluations. Journal of Business Research 67, 1209–1217.

de Ferran, F., Grunert, K.G., 2007. French fair trade coffee buyers' purchasing motives: an exploratory study using means-end chains analysis. Food Quality and Preference 18, 218–229.

Dean, M., Lampila, P., Shepherd, R., Arvola, A., Saba, A., Vassallo, M., Lahteenmaki, L., 2012. Perceived relevance and foods with health-related claims. Food Quality and Preference 24, 129–135.

European Commission, 2012. Food Information Schemes, Labelling and Logos, Internal Document DG SANCO.

Fernqvist, F., Ekelund, L., 2014. Credence and the effect on consumer liking of food–a review. Food Quality and Preference 32, 340–353.

Graham, D.J., Orquin, J.L., Visschers, V.H., 2012. Eye tracking and nutrition label use: a review of the literature and recommendations for label enhancement. Food Policy 37, 378–382.

Groeppel-Klein, A., 2011. In-store Use of Nutrition Labels (Final workshop of the FLABEL project, Brussels).

Grunert, K.G., 2005. Food quality and safety: consumer perception and demand. European Review of Agricultural Economics 32, 369–391.

Grunert, K.G., Bolton, L.E., Raats, M.M., 2012a. Processing and acting upon nutrition labeling on food: the state of knowledge and new directions for transformative consumer research. In: Mick, D.G., Pettigrew, S., Pechmann, C., Ozanne, J.L. (Eds.), Transformative Consumer Research for Personal and Collective Well-being. Routledge, New York, pp. 333–351.

Grunert, K.G., Fernández Celemín, L., Storcksdieck genannt Bonsmann, S., Wills, J.M., 2012b. Motivation and attention are the major bottlenecks in nutrition labelling. Food Science and Technology 26, 19–21.

Grunert, K.G., Fernández-Celemín, L., Wills, J.M., genannt Bonsmann, S.S., Nureeva, L., 2010. Use and understanding of nutrition information on food labels in six European countries. Journal of Public Health 18, 261–277.

Grunert, K.G., Hieke, S., Wills, J., 2014. Sustainability labels on food products: consumer motivation, understanding and use. Food Policy 44, 177–189.

Grunert, K.G., Lähteenmäki, L., Boztug, Y., Martinsdóttir, E., Ueland, Ø., Åström, A., Lampila, P., 2009. Perception of health claims among Nordic consumers. Journal of Consumer Policy 32, 269–287.

Grunert, K.G., Wills, J.M., 2007. A review of European research on consumer response to nutrition information on food labels. Journal of Public Health 15, 385–399.

Hieke, S., Kuljanic, N., Wills, J.M., Pravst, I., Kaur, A., Raats, M.M., Grunert, K.G., 2015. The role of health-related claims and health-related symbols in consumer behaviour: design and conceptual framework of the CLYMBOL project and initial results. Nutrition Bulletin 40, 66–72.

Hodgkins, C.E., Raats, M.M., Fife-Schaw, C., Peacock, M., Gröppel-Klein, A., Koenigstorfer, J., Grunert, K.G., 2015. Guiding healthier food choice: systematic comparison of four front-of-pack labelling systems and their effect on judgements of product healthiness. British Journal of Nutrition 113, 1652–1663.

Lähteenmäki, L., Lampila, P., Grunert, K., Boztug, Y., Ueland, Ø., Åström, A., Martinsdóttir, E., 2010. Impact of health-related claims on the perception of other product attributes. Food Policy 35, 230–239.

Malam, S., Clegg, S., Kirwan, S., et al., 2009. Comprehension and Use of UK Nutrition Signpost Labelling Schemes. Food Standard Agency, London. http://www.food.gov.uk/multimedia/pdfs/pmpreport.pdf.

Mazzocchi, M., Traill, B., Shogren, J.F., 2009. Fat Economics. Oxford University Press, Oxford.

Mennell, S., Murcott, A., Van-Otterloo, A.H., 1992. The Sociology of Eating, Diet and Culture. Sage, London.

Moorman, C., 1990. The effects of stimulus and consumer characteristics on the utilization of nutrition information. Journal of Consumer Research 17, 362–374.

Raats, M.M., Hieke, S., Jola, C., Hodgkins, C., Kennedy, J., Wills, J., 2015. Reference amounts utilised in front of package nutrition labelling; impact on product healthfulness evaluations. European Journal of Clinical Nutrition 69, 619–625.

Roe, B., Levy, A.S., Derby, B.M., 1999. The impact of health claims on consumer search and product evaluation outcomes: results from FDA experimental data. Journal of Public Policy and Marketing 18, 89–105.

Sacks, G., Rayner, M., Swinburn, B., 2009. Impact of front-of-pack 'traffic-light' nutrition labelling on consumer food purchases in the UK. Health Promotion International 24, 344–352.

Shi, S.W., Zhang, J., 2014. Usage experience with decision aids and evolution of online purchase behavior. Marketing Science 33, 871–882.

Sutherland, L.A., Kaley, L.A., Fischer, L., 2010. Guiding stars: the effect of a nutrition navigation program on consumer purchases at the supermarket. The American Journal of Clinical Nutrition 91, 1090S–1094S.

Thøgersen, J., Jørgensen, A.K., Sandager, S., 2012. Consumer decision making regarding a "green" everyday product. Psychology and Marketing 29, 187–197.

Wills, J.M., Storcksdieck genannt Bonsmann, S., Kolka, M., Grunert, K.G., 2012. European consumers and health claims: attitudes, understanding and purchasing behaviour. Proceedings of the Nutrition Society 71, 229–236.

Zanoli, R., Naspetti, S., 2002. Consumer motivations in the purchase of organic food: a means-end approach. British Food Journal 104, 643–653.

Designing Inclusive Packaging

3

J. Goodman-Deane, S. Waller, M. Bradley
University of Cambridge, Cambridge, United Kingdom

A. Yoxall
Sheffield Hallam University, Sheffield, United Kingdom

D. Wiggins
DRW Packaging Consultants, Leicester, United Kingdom

P.J. Clarkson
University of Cambridge, Cambridge, United Kingdom

3.1 Noninclusive Packaging

Ideally, the packaging experience would be satisfying, bringing pleasure and even delight. In practice, however, this is often not the case. Many items of packaging have features that are hard to see, manipulate, or understand. As a result, many users find it difficult or even impossible to open, read, or otherwise use the packaging (see Fig. 3.1). This particularly affects people with reduced capabilities, such as older people and those with disabilities. However, it can also impact more mainstream users, producing feelings of frustration and annoyance rather than delight. In 2004, the term "wrap rage" was coined to describe this sense of frustration about accessing and using difficult-to-open packaging on even very simple items.

Packaging must fulfill a range of roles. It has to protect and preserve the contents both in storage and transportation; it has to market and sell the product, identify the contents, and provide cooking and storage instructions; and of course it has to facilitate access to those contents.

Those roles can often be in competition and can create a difficult balancing act for designers, packer-fillers, and distributors. Legal requirements on product information on packaging can often lead to very small font sizes, and sealing forces can often be relatively high to facilitate safe transportation or prevent theft. Pressure to meet environmental legislation and hence to reduce the amount of packaging can also be seen to compete against the desire for packaging that is easy to open. For example, removing ring-pulls on food cans reduces packaging but makes the cans harder to open.

The ability to access packaging is affected by a consumer's strength and dexterity, while their understanding of the access features and of information on the packaging (such as nutritional information and use-by dates) is affected by visual acuity. All of these user capabilities are known to decline with age, with significant reductions in strength for women over 70 (Langley et al., 2005; Yoxall et al., 2008), dexterity declining by approximately 1.6% a year over the age of 60 (Desrosiers et al., 1993),

Integrating the Packaging and Product Experience in Food and Beverages. http://dx.doi.org/10.1016/B978-0-08-100356-5.00003-6

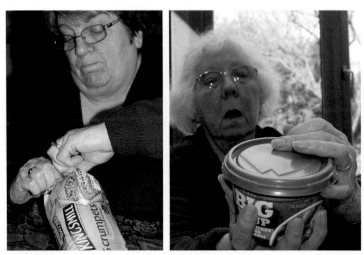

Figure 3.1 Users struggling to open packaging and read cooking instructions on packaging.

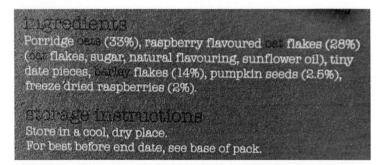

Figure 3.2 Example of an ingredients list where an attempt has been made to highlight allergens (oats and barley) by printing them in darker text. However, this has rendered the information difficult to read due to poor contrast with the background.

and age-related long sightedness affecting around 83% of the population aged over 45 (Holden et al., 2008).

The context of use (ie, where and when consumers use packaging) can also affect the ability to access the pack for people with no significant capability loss. For example, wet or slippery hands can make handling packaging difficult and poor lighting conditions can make reading pack information difficult even for people with average eyesight.

In particular, allergen advice means that problematic ingredients are labeled in bold. However, poor use of color and contrast choice can mean that finding and reading this information can be difficult (see Fig. 3.2). Fig. 3.3 shows a further two examples where participants in tests run by the authors failed to identify access or product information correctly.

Several surveys have listed items that consumers have had significant issues with, in terms of accessing the packaging. For example, *Yours* magazine (McConnell, 2004)

 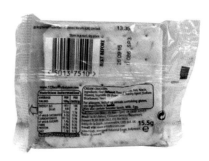

Figure 3.3 Examples of packaging for which users under tests failed to identify product or packaging accessibility information.

listed the following as difficult items for people to access: jam jars, shrink-wrapped cheese, tins of meat and fish, medicines, and child-resistant closures on bleach. A separate survey by the packaging specialist Payne added plastic-molded clamshells to the list (Packaging Europe, 2013).

In fact, a 2013 survey found that two-thirds of consumers "get frustrated trying to get into everyday packaging" and "four in ten people have hurt themselves while trying to open packaging over the last two years" (Which? Magazine, 2013). Access to packaging is often seen as a trivial issue that can be overcome with scissors or other implements such as knives, hot water or towels, or by asking for help from relatives (Yoxall et al., 2010). However, research has identified individuals living on ready meals, waiting for visitors to help open packaging, and changing purchasing behavior because of difficulties with packaging (Yoxall, 2013). Perhaps most seriously, access to packaging was seen to lead to nutritional problems in hospitals in the New South Wales region of Australia (Bell et al., 2013).

Issues to do with the use of packaging extend beyond accessing the contents to problems with purchase, delivery, storage, dispensing, reuse, and disposal. For example, frustration and difficulty can arise when trying to recognize a particular product in a shop, pour from a carton of milk, reclose a resealable packet, check whether packaging is recyclable, or flatten a box to dispose of it.

Some examples of "easy-open" packaging have recently appeared, and recent CEN and ISO standards help to encourage and enable the development of more accessible packaging (Great Britain, British Standards Institution, 2011; International Organisation for Standardization, 2015). However, there are still significant issues with many packaging formats. This problematic packaging can be termed noninclusive: packaging that does not include a wide range of people, ages, and abilities and results in frustration, difficulty, and exclusion. The issues posed by such packaging are becoming an increasing problem as the population ages and increasing numbers of people are affected by capability loss. Twenty-three percent of the UK population are now aged 60 or over, and over 18% of the population have at least one moderate or severe capability loss (Waller et al., 2010).

3.2   Inclusive Design

Inclusive design is one way to address these challenges (Keates and Clarkson, 2003; Waller et al., 2015). It can be defined as "The design of mainstream products and/or services that are accessible to, and usable by, as many people as reasonably possible … without the need for special adaptation or specialised design." (Great Britain, British Standards Institution, 2005)

Thus, inclusive design applies to standard packaging. It is not the design of specialized packaging for older people or specialized markets. It is about making everyday packaging easier to open and use. Someone with arthritis would be able to open an inclusively designed jar without having to buy a special implement to attach to the lid. An older person would be able to read the instructions on the side of a packet without having to get out a magnifying glass. As a side effect, people with average eyesight would also be able to read the instructions if the lighting in their kitchen was dim.

However, inclusive design also recognizes that there are limits to this approach. It may not be practically possible to make an item of packaging accessible to the entire population. For example, it may not be possible to make a jar lid easy enough to turn for someone with an extremely weak grip, while still providing a sufficient seal to keep the food fresh. Similarly, there may not be enough space on the packaging to print instructions large enough for someone with severe vision impairments to be able to read them.

Inclusive design is about making informed decisions based on an understanding of the target market. It seeks to extend the number of people who can use the packaging, but it also recognizes a range of other constraints and success criteria, such as cost and technical viability (see Section 3.3.1). Inclusive design seeks to maximize the packaging experience, bearing in mind these other constraints (Waller et al., 2015).

Inclusive design is also related to "universal design" and "design for all," which similarly seek to broaden the range of people who can use mainstream products (Preiser and Ostroff, 2001). However, inclusive design typically places more emphasis on informing commercial decisions as described above.

Inclusive design provides a framework within which specific evaluation methods and accessibility techniques can be used, such as those recommended in the recent CEN and ISO standards (Great Britain, British Standards Institution, 2011; International Organisation for Standardization, 2015).

3.3   A Framework for Inclusive Design

For inclusive design to be put into practice effectively, it needs to be part of the design team's thinking from the start of the design process, not added as an extra at the end. Last-minute modifications tend to be both very expensive and not very effective (Mynott et al., 1994). If inclusivity is considered from the start, more innovative and effective solutions can be considered. Thus, while specific tools for inclusive design are valuable (and will be discussed later in this chapter), they are not

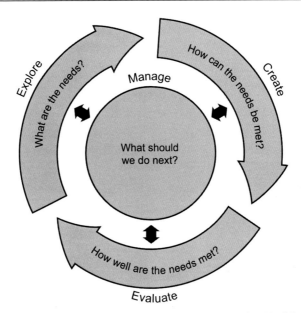

Figure 3.4 A model for the inclusive design process (Inclusive Design Toolkit, 2015a).

enough—incorporating inclusive design principles into the design process as a whole is important for effective inclusive design.

We therefore propose a model of a concept design process that allows inclusive design thinking and principles to be integrated throughout. The model is summarized in Fig. 3.4. It includes four main phases:

1. Explore: Determine "What are the needs?"
2. Create: Generate ideas to address "How can the needs be met?"
3. Evaluate: Judge and test the design concepts to determine "How well are the needs met?"
4. Manage: Review the evidence to decide "What should we do next?"

Fig. 3.4 shows how these phases fit together. Successive cycles of the Explore, Create, and Evaluate phases are used to generate a clearer understanding of the needs, better solutions to meet these needs, and stronger evidence that the needs are met. The Manage phase is used to steer and direct the design process to optimize resource utilization and to maximize the quality of output within the project constraints.

This chapter briefly describes each of these phases. More detailed information can be found on the Inclusive Design Toolkit (2015a).

3.3.1 Explore

The Explore phase is about gaining a deeper understanding of the criteria that the packaging needs to fulfill. These criteria are based on the needs of a whole range of stakeholders across the whole life cycle of the packaging. These stakeholders include users, manufacturers, retailers, and many others. The criteria in general can be

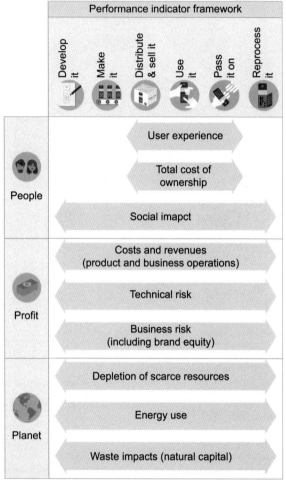

Figure 3.5 A framework for criteria that packaging needs to fulfill (Inclusive Design Toolkit, 2015a).

summarized in a "performance indicator framework" as shown in Fig. 3.5, although the particular criteria and their relative importance will vary depending on the product and packaging.

Thus the criteria may include user-oriented aspects such as ease of opening, instructions for use, ability to know what allergens are in the product, and aesthetics. However, they also include aspects such as branding, recyclability, space-efficient warehouse storage, and cost of manufacture. Inclusive design recognizes that packaging design is done within a context of multiple criteria, and that inclusivity is not the only concern. This practicality often results in more workable solutions in practice.

During the Explore phase, it is also useful to identify "dealbreaker" issues, ie, concerns of sufficient importance to one or more stakeholders that the concept cannot go ahead if they are not addressed adequately. Examples of these include issues to do with legal compliance, as well as brand issues which may enforce the use of certain colors or the placement of images or logos in particular places.

In summary, the Explore phase is about understanding the criteria better. This involves exploring the needs and desires of the different stakeholders, and understanding how the packaging will be used throughout the whole user life cycle from purchase to disposal.

3.3.2 Create

The Create phase is about creating possible solutions to meet the needs and criteria identified by the Explore phase. This includes many of the activities commonly thought of as belonging to concept design, ranging from producing initial ideas to developing them into prototypes that can be tested.

In this phase, it is important for the design team not to get fixated on one idea or a small subset of ideas. Effective inclusive design often requires thinking more widely and exploring different and possibly unusual options. To enable this, it is important to examine a wide range of stimuli, both obvious and unexpected. It is also necessary to develop a creative culture in which people are prepared to give things a go and see what happens, even if what happens is not successful. Impractical ideas often spur realistic ones that are much better than those that could be obtained through sole consideration of the possible.

The Create phase may also involve challenging the constraints and pushing the boundaries set in the Explore phase. This can help to refine the understanding of the criteria, and can result in more innovative and effective solutions, even if the criteria remain unchanged.

3.3.3 Evaluate

The Evaluate phase examines the concepts to determine how well they meet the criteria identified earlier, both in terms of the performance indicators and dealbreaker issues. This is vitally important to ensure that the criteria are actually met. Without this, the team runs the danger of just choosing concepts that they like the look of and that work for them, but not based on the wider target population or key stakeholders. The evaluation process not only checks for such issues, but can identify ways to refine concepts to solve problems and improve performance.

Evaluation is best done early in the design process while meaningful change is still possible. To enable this, it is often necessary to perform quick tests with rough prototypes, rather than waiting until full prototypes are ready.

There are many different ways of evaluating concepts, including testing with users, expert appraisal, and exclusion estimation (Goodman-Deane et al., 2014). In particular, the CEN15945 Technical Specification (Great Britain, British Standards Institution, 2011) encourages the evaluation of a pack or pack concept against a defined user test panel, the demographic of that panel being determined by the specification. Doing this can establish a degree of understanding and evaluation of pack performance. The standard has been successfully used across a range of packaging formats, modified following extensive use, and included in ISO17480, *Packaging – Accessible design – Ease of opening* (International Organisation for Standardisation, 2015). It can be particularly useful if used in conjunction with observation and interview techniques.

However, user testing can be expensive and time-consuming and may not always be feasible, particularly for quick iteration of early concepts. A range of evaluation methods used throughout the design process can be very effective.

3.3.4 Manage

The inclusive design process is dynamic and iterative, with the team moving between different phases in response to the needs of the project and the outcomes of the steps. For example, the Evaluate phase may identify some changes that need to be made to the concepts as well as suggestions for improvements, which trigger a move back to the Create phase. Alternatively, the Evaluate phase may uncover a lack of understanding of some of the user needs, which necessitates further Exploration.

The Manage phase manages this whole process, keeping the project on course and on budget. It determines what needs to be done next and when to move on to the next stage. In order to do this effectively, it is important for the team to take a step back periodically to review its progress and refine its goals. Maintaining good communication between project partners and keeping an eye on the business case are also key to ensuring a successful project.

3.4 Empathy Tools

There are many tools and activities which are helpful in supporting inclusive design. However, there is not enough space in this chapter to describe them all. Instead, we focus on a particular subset of tools, sometimes called empathy tools. Information about other tools can be found in the Inclusive Design Toolkit (2015a).

Empathy tools are intended to help develop empathy with and understanding of users. This is important because, as Cooper (1999) points out, designers tend to design for people with similar capabilities and skills to themselves in the absence of specific prompts to do otherwise. This may work if there is a restricted target user group that is very similar to the designer, eg, when software designers produce a development system for other software designers to use. However, in packaging design and inclusive packaging design in particular, the target user group is generally larger and more diverse.

Empathy tools help to break designers out of the mind-set of designing for themselves, making them think instead about the needs of a wider range of users. Empathy can be a powerful motivator for designers and provide useful insight for understanding and improving users' packaging experiences.

These tools are useful throughout the design process (described in Section 3.3). They can help the design team to understand user needs and provide some user perspective in an initial quick evaluation of concepts. They can also be a great help in communicating with clients and convincing them of the value of inclusive design proposals.

Empathy and understanding may be most effectively built through direct in-depth involvement of real users in the design process. However, in many cases this is not practical, requiring substantial cost and training to do properly. In other cases, users may be involved in a project, but only at particular points. The aim of the empathy tools described in this section is to bridge this gap, providing quick, easy-to-use, low-cost methods to enable quick insights and feedback when users are not available. Please note that they are not intended to replace the need to involve real users, and are most effectively used alongside user involvement as part of a holistic inclusive design process.

In addition, these methods do some things that user involvement cannot do, at least not without large numbers of users. They can incorporate and summarize the situations of many different users with different levels of capabilities.

This chapter describes two empathy tools in particular: simulation and personas.

3.5   Simulation

In simulation methods, designers are given a first-hand experience of some of the functional effects of capability loss (Cardoso and Clarkson, 2006). This can be done by wearing special equipment that reduces their capabilities or by using software that shows what packaging might look like to someone with a vision impairment.

Note that when using simulation, it is important to remember that it only provides a limited experience of what a disability is like. It does not convey the frustration, difficulties, social consequences, or coping strategies associated with living with an impairment long term. It also does not usually include pain and other symptoms associated with the impairment. As a result, it does not replace user involvement, but supplements it, helping a designer to internalize information obtained through other methods and providing initial feedback before designs are taken to users.

3.5.1   Wearable Simulators

Wearable simulators directly restrict the wearers' capabilities so that they find it more difficult to see, hear, or move. These include full body restrictors like the Third Age Suit (Hitchcock et al., 2001) and items that limit particular abilities. Full body restrictors are expensive and difficult to get hold of. Therefore, when examining packaging, it is generally more useful to use simulators that limit the abilities most relevant to packaging use, such as vision and dexterity. Cognitive ability is also very important to packaging use, but at present effective and practical cognitive loss simulation is not available.

3.5.1.1   Vision Impairment Simulators

Vision impairment is commonly simulated by wearing glasses that obscure one's vision. The accuracy of this simulation widely varies. At one end of the spectrum are very rough methods such as smearing petroleum jelly on glasses to obscure one's vision (Nicolle and Maguire, 2003). More accurate methods typically use glasses that have been designed to simulate particular levels and types of vision loss, such as VINE Sim Specs (Visual Impairment North East, n.d.) or the Low Vision Simulators provided by Fork in the Road (2015). However, many of these are not designed for use in inclusive design and often do not cover the milder degrees of vision loss of most relevance to inclusive packaging design.

The Cambridge Simulation Glasses, produced by the University of Cambridge, have been designed for use in inclusive design (Inclusive Design Toolkit, 2015b). Each pair of glasses simulates a mild loss of visual acuity and they can be layered up to simulate higher degrees of vision loss (see Fig. 3.6). The effect of different numbers of glasses on the wearer's visual acuity has been measured (Goodman-Deane et al., 2013). Therefore, they can be used both to increase general empathy with those with mild vision loss and to simulate specific degrees of visual impairment.

The glasses can be used to examine the visual accessibility of packaging. A recommended procedure for doing this is described in Goodman-Deane et al. (2014) and summarized in Fig. 3.7. Each assessor's eyesight is first estimated using a simplified

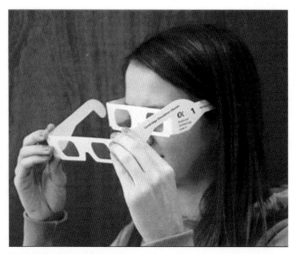

Figure 3.6 The Cambridge Simulation Glasses.

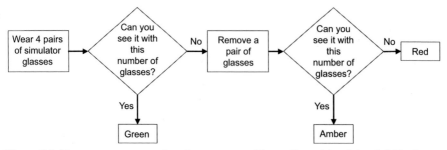

Figure 3.7 The assessment procedure for assessors with excellent vision (visual ability better than 20/16).

vision chart. Depending on their eyesight, each assessor is given a particular number of simulator glasses to wear. The assessors then examine the packaging while wearing the given number of glasses, focusing on features for which visual accessibility is important, such as the visibility of easy-open tabs or the legibility of ingredients lists. It is important that assessors keep the product at a normal working distance while doing this, to avoid skewing the results. They determine whether they can see the features well enough to be able to use them (eg, recognize the tabs or read the ingredients list). Depending on the number of glasses with which they can do this, the features are rated green, amber, or red.

Green indicates that less than 1% of the population would be excluded from using that packaging feature on grounds of visual acuity. Amber indicates that between 1% and 6% of the population are excluded, and Red indicates that more than 6% would be excluded (details of the calculations can be found in Goodman-Deane et al., 2014). These figures only relate to exclusion on the grounds of visual acuity. Of course, there may be other vision problems with the packaging—eg, those associated with tunnel vision or color blindness. Ideally, this procedure should be used together with user involvement as part of an overall inclusive design process. Nevertheless, this procedure can perform a useful sanity check and uncover significant issues with the packaging at an early stage of the design process.

3.5.1.2 Dexterity Impairment Simulators

Wearable simulators that restrict hand function include rough methods like taping coins to the back of one's knuckles and wearing gardening gloves (Nicolle and Maguire, 2003). Gloves produced specifically for simulation purposes offer a more consistent and reproducible simulation experience, which is useful when assessing packaging options. Commercially available examples include Georgia Tech's Arthritis Simulation Gloves (Georgia Tech, n.d.) and the Cambridge Simulation Gloves (Inclusive Design Toolkit, 2015b) shown in Fig. 3.8. The Arthritis Simulation Gloves have been used by Arthritis Australia for educational purposes and by various

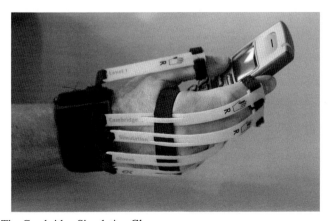

Figure 3.8 The Cambridge Simulation Gloves.

companies to examine the difficulties that people with arthritis might have in opening and using products (The Engineer, 2010). The Cambridge Simulation Gloves have been used by Age UK as well as various companies, as illustrated in the case studies below.

Unlike the glasses, the particular levels of capability loss simulated by the gloves for a "typical" wearer have not been measured. This is because hand strength and size varies dramatically, even among people considered "fully able." Furthermore, the effect of the gloves is sensitive to fit and minor adjustments made in putting them on. As a result, the effects of the gloves will be different for every person who wears them. They are therefore generally not appropriate for use in a pass/fail evaluation as was described for the glasses (Fig. 3.7).

Nevertheless, the gloves are useful in providing empathy and insight into the effects of dexterity loss on product use. The harder a product is to use while wearing the gloves, the more demand it places on dexterity and the more inaccessible it is. In particular, the Cambridge Simulation Gloves limit the strength and range of motion of the fingers and thumb—they affect the ability to bend one's fingers. Bending fingers is extremely painful for people with arthritis in their knuckle joints, and tasks that do not require the hand to make intricate shapes are therefore more inclusive.

3.5.2   Software Simulators

Another route for simulation is provided by software. Image or sound files can be manipulated by a computer to give the user an impression of what they might look or sound like to someone with a vision or hearing impairment, respectively. The former is usually the most relevant in the context of packaging.

There are several vision simulators available. Some specialize in particular impairments such as color blindness (eg, Vischeck, n.d.). Others simulate a range of impairments such as macular degeneration, glaucoma, and cataract. Examples include VIS (University of Illinois, 2005) and the VisionSim app produced by The Braille Institute (2013). These are usually intended for either general education purposes or for use by software designers (eg, web designers).

The Impairment Simulation Software produced by the University of Cambridge (Goodman-Deane et al., 2007) is particularly intended for use in product and graphic design. The screenshot in Fig. 3.9 shows a simulation of a mild level of macular degeneration applied to the packaging example in Fig. 3.2. Designers can upload images of existing packaging or proposed concepts, and see the effects of various vision conditions including macular degeneration, diabetic retinopathy, glaucoma and cataracts. These conditions are simulated at a range of degrees of severity. Red/green color blindness is also simulated. Alternatively, the user can explore the effects of the loss of various visual functions, such as visual field, contrast sensitivity, and visual acuity.

Software simulations have some advantages over wearable simulators. They allow a designer to explore the effects of a wider range of impairments and to adjust the degree of severity of the impairments more easily. They are also useful for examining digital artwork before it exists in reality. A particularly effective use of them is to create images to embed within a presentation or report, to illustrate an inclusivity

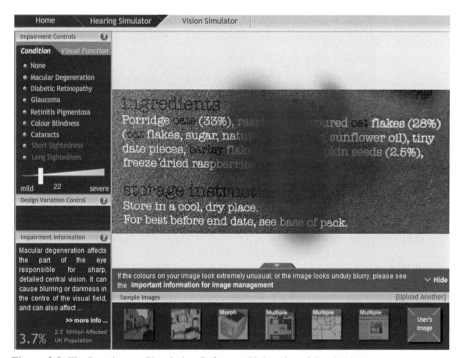

Figure 3.9 The Impairment Simulation Software (University of Cambridge).

issue or to show that such an issue has been addressed effectively. However, they do have disadvantages. In particular, there can be issues to do with the clarity of image reproduction and with apparent size and lighting that can reduce the accuracy of the simulation and its realism for real-world outcomes.

In addition, wearable simulators have some advantages. In particular, they provide the designer with a much more direct experience, which can be more effective in building empathy and insight. In addition, wearable simulators allow simulation to be applied while directly manipulating the packaging in context. This allows the designer to experience how vision impairment would affect the whole packaging experience, and allows vision and dexterity simulation to be combined.

3.5.3   Case Study 1: Biscuit Packaging

The Cambridge Simulation Gloves and Glasses were used by the authors to compare the inclusivity of current and proposed packaging for Family Circle boxes for United Biscuits. The assessors first performed a task analysis of the packaging from purchase to disposal. They then worked through each of these tasks, examining the demands the packaging placed on the users' capabilities. Simulation gloves were used to help assess the level of demand placed on dexterity, and glasses to assess the demand placed on vision.

For example, in order to open one version of the packaging, the user first had to remove a cardboard sleeve (see Fig. 3.10). The intention was that the user would

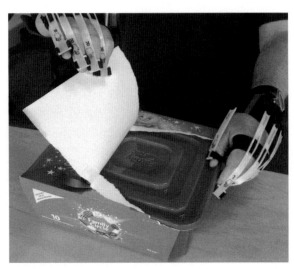

Figure 3.10 Assessment of proposed biscuit packaging.

slide the sleeve off the box. However, attempting to do this while wearing the simulation gloves was difficult. It required fairly high force and a good grip, excluding many users. Some users may have to resort to tearing the sleeve in order to remove it (as shown in Fig. 3.10), because this required lower levels of dexterity. Doing so successfully completed the task of removing the sleeve, but had the side effect of damaging the sleeve, which contained the ingredients list and information on the different types of biscuits in the box.

In a similar manner, the simulation glasses were used to examine visual tasks associated with the packaging, such as checking the best-before date, checking tamper evidence, determining how to open the box (including finding the opening tab), and checking for allergens.

While the glasses were used to identify general issues with the packaging, the procedure described in Section 3.5.1.1 had not yet been developed at the time of the study. However, it was applied to the packaging at a later date as a proof of concept, as described in Goodman-Deane et al. (2014). The procedure was used to examine the ingredients list to check for allergens. The assessor had excellent vision and started with four pairs of simulator glasses, as shown in Fig. 3.7. She could not complete the task with four pairs of glasses, but could do so with three pairs, resulting in the ingredients list being rated amber, corresponding to between 1% and 6% of users being excluded from this task on visual acuity grounds alone. In practice, the thinking demands associated with locating and reading the information would probably increase exclusion to above 6%.

The glasses could also be useful in determining how to improve the legibility of the ingredients information. Different options could be tried, such as using a bolder font or increasing contrast, and their effectiveness examined using the glasses.

As a result of the evaluation, the biscuit packaging was reviewed and changes were made to the proposed packaging. Learnings will be built into future design briefs.

Figure 3.11 Assessment of "easy-open" sausage packaging.

3.5.4   Case Study 2: Resealable Labels

Snack food such as cocktail sausages and sausage rolls and fruit items such as straw-berries are typically packed in clear plastic punnets (using recycled polyethylene terephthalate) with thin film lidding (Mylar). A desire to increase product shelf life and storage of the contents has led to the use of resealable peelable labels on some of these products. This presented an opportunity for McFarlane labels to assess whether this type of accessibility feature could improve the "openability" experience for consumers along with the other advantages listed. Hence, a two-stage study was commissioned and undertaken by Sheffield Hallam University.

The first stage was a consumer packaging test as specified in CEN1594 (Great Britain, British Standards Institution, 2011). This test takes a minimum of 20 older participants (the makeup of which is defined by the CEN standard). Participants are asked to familiarize themselves with the packaging. They subsequently open the pack-aging and rank their experience on a Likert scale represented by a series of smiley faces (as shown in Fig. 3.11).

Six different pack formats were tested, two without any resealable labels and four with labels from different brand owners and retailers. A pack was considered a fail if any of the 20 participants were unable to open it.

The second stage of the study used the Cambridge Simulation Gloves with a younger cohort. The group opened the packaging while wearing the gloves and also ranked the packaging using a Likert scale.

This study facilitated several observations in terms of both the pack accessibility and the use of simulation gloves versus "real" older people. First, resealable packaging was seen to score more highly on the Likert scale for both cohorts, indicating a higher desirability and usefulness. More importantly, while both examples of nonresealable packaging failed the CEN15945 test, only one example of the resealable pack created any problems, and this was due to users being unable to identify the location of the tab as it had been incorporated into the branding. This highlighted the consistent tension between the demands placed on packaging outlined earlier (to create a shelf presence and a consistent brand) and the desire to facilitate easy opening. Indeed, the pack with the most obvious label and instructions performed best in the test.

In comparing the cohorts, it was clear that the gloves enabled the younger cohort to experience similar problems to those of the older cohort, including needing to use their teeth to successfully access the pack (see Fig. 3.11) and developing a sense of frustration at being unable to open packs.

3.6 Personas

Personas are another tool which can help to build designers' empathy with and encourage them to consider the needs of a wider range of users. Personas are descriptions of fictional users that represent the target users, encapsulating data on their abilities, lifestyles, needs, and wants. They were first proposed by Cooper (1999) as a practical design tool. Ideally they should be developed specifically for each project to summarize and bring to life the user research for that project. However, in the context of inclusive design, even a generic set of inclusive design personas can be effective in encouraging designers to think more widely about their target user group.

An example set of inclusive design personas is provided in the Inclusive Design Toolkit (2015b) and shown in Fig. 3.12. This set was constructed as a training

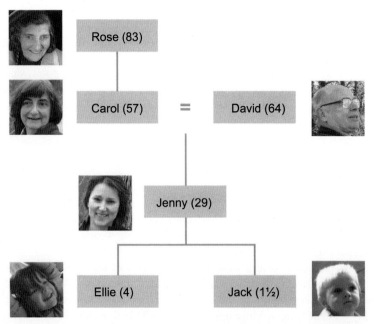

Figure 3.12 An example set of personas which covers a range of ages, life situations, and capabilities.

tool. The personas cover a range of ages, situations, and capabilities. They include Rose, who at the age of 83 struggles with everyday tasks and has difficulties with vision and hearing. Rose represents a key demographic of people who might be given a box of chocolates or biscuits as a present. However, the persona set also highlights inclusive design issues associated with less obvious cases, such as the difficulties experienced by Jenny, a young mother with her arms full of shopping bags and distracted by children. It also highlights issues experienced by people like David, aged 64, who is generally fit and healthy but experiencing some of the problems with vision that naturally come at that age. He struggles to accept these difficulties and would greatly resist any implication that he is "disabled" or needs particular help.

Personas can be used in various ways. Even simply presenting them to the design team can be useful in stimulating interest and discussion of inclusive design issues. They can also be used in examining packaging concepts. The designers can be encouraged to consider how different personas would respond to or cope with the different concepts. This does not give a rigorous assessment of accessibility, but can usefully broaden the discussion. It can help to break the fixation on one's own experience that often arises, where designers refuse to accept that a concept is noninclusive on the basis that they themselves or people they know would be able to use it without difficulty.

The use of personas can be combined with simulators by simulating the capabilities of some of the personas using gloves or glasses. This can provide a less subjective examination of whether a particular persona would be able to open a piece of packaging or not.

3.6.1   Case Study 3: Black Magic

The personas in Fig. 3.12 and the Cambridge Simulation Gloves and Glasses were used together with user trials and interviews to assess and redesign packaging for boxes of Black Magic chocolates for Nestlé.

The design team then used the scales in Fig. 3.13 to consider how each of the personas might find the packaging in terms of usability, aesthetics, and user experience. For example, they rated the usability of the packaging for each persona on a scale from "easy" to "impossible."

The design team used the simulators to help them determine these ratings. In particular, two of the personas (David and Rose) had limitations in their vision. The design team simulated this using the simulation glasses. Rose's eyesight was simulated with four pairs of glasses, and David's with two. To determine how David and Rose would experience the packaging, the designers put on the appropriate pairs of simulation glasses and viewed and tried to open the packaging themselves.

As a result of this evaluation, the design team reframed their perceptions of the target market for the product, and identified that significant improvements to the packaging were readily achievable. Several months later, the improved product made it to market with better contrast, a clearer font, and a shallower tray.

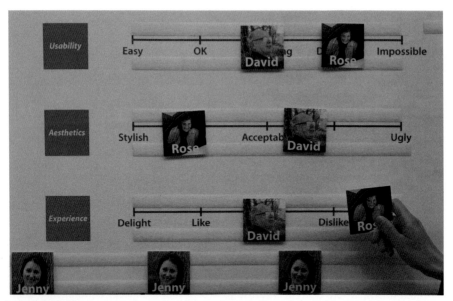

Figure 3.13 The personas were placed on scales to indicate how the designers thought they would judge the usability, aesthetics, and experience associated with the range of tasks for opening, consuming, and disposing of a box of chocolates. The scales were also used to compare Black Magic against competitors.

3.7 Conclusions

This chapter has argued for the importance of considering a wide range of users in packaging design, including those who are older or have capability limitations. Including such users often leads to improvements for more mainstream users as well, reducing frustration and difficulty and improving the packaging experience.

Inclusive design is a key way to address these issues. For inclusive design to be put into practice effectively, it needs to be considered throughout the design process. This chapter has proposed an inclusive design methodology, where different phases of the iterative process explore the needs, create possible solutions, evaluate the concepts, and manage the process.

There are many tools and methods that can be used to support inclusive design. This chapter has focused on empathy tools, which are intended to help develop designers' empathy with and understanding of users. Two such tools have been presented in more detail: capability loss simulators and personas. Simulators enable designers to experience some of the functional effects of capability loss for themselves, while personas encapsulate some of the characteristics and experiences of a range of users, bringing them to the forefront of designer awareness throughout the design process.

Putting these principles into practice and using tools like these can result in packaging that is attractive, is easy to use, and provides a positive and engaging packaging experience for a wide variety of users.

Inclusive packaging is likely to become more of an issue in the future, as the population ages and there is an increasing drive toward enabling independent living and reducing the cost of social care. While the sustainability agenda is likely to dominate packaging design, true sustainability involves the interlinking of environmental, economic, and social sustainability, of which inclusive design is a part.

3.8   Future Work and Trends

As the case studies in this chapter indicate, along with the development of guidelines and technical specifications, there is some movement toward developing more inclusively designed packaging, which is likely to increase. A shopping trip to any major retailer will identify examples of packaging labeled "easy-open," including household items such as toothbrushes and razors and foodstuff such as jam jars, pasta, cereals, and cheese (see example in Fig. 3.14).

Research work in assessing packaging is ongoing, with recent studies looking at dexterity, affordances, and legibility (eg, Rowson et al., 2014; de la Fuente et al., 2015). In a survey for *Packaging News* magazine, while cost reduction and shelf impact took top spots for key drivers in 2014, quality enhancement and openability/convenience were also trends noted in the survey (Chadwick, 2014).

There is also ongoing work on developing methods and tools for inclusive design. In particular, recent work on simulator methods has examined how to make them fit better with the working practices of designers (eg, Cornish et al., 2014).

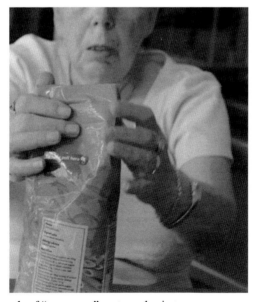

Figure 3.14 An example of "easy-open" pasta packaging.

References

Bell, A.F., Walton, K., Chevis, J.S., Manson, C., Wypch, A., Yoxall, A., Kirkby, J., Alexander, N., 2013. Accessing packaged food and beverages in hospital. Exploring experiences of patients and staff. Appetite 60 (1), 231–238.

Cardoso, C., Clarkson, P.J., 2006. Impairing designers: using calibrated physical restrainers to empathise with users. In: The 2nd International Conference for Universal Design in Kyoto. Kyoto, 22–25 Oct 2006. International Association for Universal Design.

Chadwick, P., 2014. Cost Is the Innovation Driver, Says Survey | Easyfairs/Packaging News Buying Survey Results. Packaging News (Online) 6 August. Available from: http://www. packagingnews.co.uk/features/cost-is-the-innovation-driver-says-survey-easyfairspackaging-news-buying-survey-results-06-08-2014 (accessed 28.09.15.).

Cooper, 1999. The Inmates Are Running the Asylum. SAMS Publishing.

Cornish, K., Goodman-Deane, J., Clarkson, P.J., 2014. Designer requirements for visual capability loss simulator tools: differences between design disciplines. In: UAHCI 2014, HCI International 2014. Crete, Greece. 22–27 June 2014.

Desrosiers, J., Hébert, R., Dutil, E., Bravo, G., 1993. Development and reliability of an upper extremity function test for the elderly: the TEMPA. Canadian Journal of Occupational Therapy 60 (1), 9–16.

de la Fuente, J., Gustafson, S., Twomey, C., Bix, L., 2015. An affordance-based methodology for package design. Packaging Technology and Science 28 (2), 157–171.

Fork in the Road, 2015. Fork in the Road: Vision Rehabilitation Services LLC. (Online) Available from: http://www.lowvisionsimulators.com (accessed 18.09.15.).

Georgia Tech Research Institute, n.d. Arthritis Simulation Gloves. (Online) Available from: http://hseb.gtri.gatech.edu/gloves.php (accessed 27.08.15.).

Goodman-Deane, J., Langdon, P.M., Clarkson, P.J., Caldwell, N.H.M., Sarhan, A.M., 2007. Equipping designers by simulating the effects of visual and hearing impairments. In: Assets 2007. Tempe, 14–17 Oct 2007. ACM Press, pp. 241–242.

Goodman-Deane, J., Waller, S., Collins, A.-C., Clarkson, P.J., 2013. Simulating vision loss: what levels of impairment are actually represented? In: Ergonomics & Human Factors 2013. Cambridge, 15–18 April 2013. Institute of Ergonomics & Human Factors.

Goodman-Deane, J., Waller, S., Cornish, K., Clarkson, P.J., 2014. A simple procedure for using vision impairment simulators to assess the visual clarity of product features. In: UAHCI 2014, HCI International 2014, Part I. Crete, 22–27 June 2014. Springer, pp. 19–30. LNCS 8513.

Great Britain, British Standards Institution, 2005. Design Management Systems – Managing Inclusive Design – Guide. BSI, London, (Standard BS 7000-6:2005).

Great Britain, British Standards Institution, 2011. Packaging Ease of Opening: Criteria and Test Methods for Evaluating Consumer Packaging. BSI, London, (CEN-15945).

Hitchcock, D.R., Lockyer, S., Cook, S., Quigley, C., 2001. Third age usability and safety – an ergonomics contribution to design. International Journal of Human-Computer Studies 55 (4), 635–643.

Holden, B.A., Fricke, T.R., Ho, S.M., Wong, R., Schlenther, G., Cronjé, S., Burnett, A., Papas, E., Naidoo, K.S., Frick, K.D., 2008. Global vision impairment due to uncorrected presbyopia. Archives of Ophthalmology 126 (12), 1731–1739.

Inclusive Design Toolkit, 2015a. Inclusive Design Toolkit. (Online) Available from: www. inclusivedesigntoolkit.com (accessed 27.08.15.).

Inclusive Design Toolkit, 2015b. Inclusive Design Tools. (Online) Available from: www. inclusivedesigntoolkit.com/tools (accessed 27.08.15.).

International Organisation for Standardization, 2015. Packaging – Accessible Design – Ease of Opening. ISO, Geneva, (ISO 17480).

Keates, S., Clarkson, P.J., 2003. Countering Design Exclusion: An Introduction to Inclusive Design. Springer, London.

Langley, J., Janson, R., Wearn, J., Yoxall, A., 2005. 'Inclusive' design for containers: improving openability. Packaging Technology and Science 18 (6), 285–293.

McConnell, V. (Ed.), 2004. Pack It in! Just Say No to Impossible Packaging. Yours Magazine, pp. 16–18. 30 Jan–27 Feb 2004.

Mynott, C., Smith, J., Benson, J., 1994. Successful Product Development: Management Case Studies. Department of Trade and Industry. Report available from: M90s Publications, DTI, Admail 528, London SW1W 8YT.

Nicolle, C.A., Maguire, M., 2003. Empathic modelling in teaching design for all. In: Universal Access in HCI, HCI International, vol. 4, pp. 143–147.

Packaging Europe, 2013. Payne Unveils Top Consumer Frustrations. (Online) Available from: http://m.packagingeurope.com/packaging-europe-news/51701/payne-unveils-top-con-sumer-frustrations.html (accessed 27.08.15.).

Preiser, W.F.E., Ostroff, E., 2001. Universal Design Handbook. McGraw-Hill, New York, USA.

Rowson, J., Sangrar, A., Rodriguez-Falcon, E., Bell, A.F., Walton, K.A., Yoxall, A., Kamat, S.R., 2014. Rating accessibility of packaging: a medical packaging example. Packaging Technology and Science 27 (7), 577–589.

The Braille Institute, 2013. VisionSim (v3.0) [Mobile Application Software]. Available from: http://itunes.apple.com (accessed 28.09.15.).

The Engineer, 2010. Arthritis-Simulation Gloves. The Engineer (Online) 10 Feb 2010. Available from: http://www.theengineer.co.uk/news/arthritis-simulation-gloves/1000895. article (accessed 27.08.15.).

University of Illinois, 2005. Visual impairment simulator for Microsoft Windows. (Online) Available from: http://vis.cita.uiuc.edu (accessed 27.08.15.).

Vischeck, n.d. Vischeck. (Online) Available from: http://www.vischeck.com/ (accessed 27.08.15.).

Visual Impairment North East, n.d. Welcome to Vine Sim Specs. (Online) Available from: http://www.vinesimspecs.com/ (accessed 27.08.15.).

Waller, S.D., Langdon, P.M., Clarkson, P.J., 2010. Designing a more inclusive world. Journal of Integrated Care 18 (4), 19–25.

Waller, S., Bradley, M., Hosking, I., Clarkson, P.J., 2015. Making the case for inclusive design. Applied Ergonomics 46, 297–303.

Which? Magazine, 2013. Which? Lifts the Lid on Packaging that Gives You 'Wrap Rage'. (Online) Available from: http://www.which.co.uk/news/2013/08/which-lifts-the-lid-on-packaging-that-gives-you-wrap-rage-330061/ (accessed 27.08.15.).

Yoxall, A., 2013. Invited Seminar. The University of Wollongong, Australia.

Yoxall, A., Janson, R., 2008. Fact or friction: a model for understanding the openability of wide-mouth closures. Packaging Technology and Science 21, 137–147.

Yoxall, A., Langley, J., Musselwhite, C., Rodriguez-Falcon, E.M., Rowson, J., 2010. Husband, daughter, son and postman, hot water, knife and towel: assistive strategies for jar opening. Designing Inclusive Interactions: Inclusive Interactions Between People and Products in Their Contexts of Use 187–196.

Omni-Channel Retail—Challenges and Opportunities for Packaging Innovation

C. Barnes

C. Barnes
Leeds Beckett University, Leeds, United Kingdom

4.1 Introduction

Over the past two decades, the world of retailing has gone through significant change and this has had an impact on the innovation needs for both food products and their packaging. Increasing digital (both online and mobile) shopping for groceries has been partly responsible for these developments. However, social trends, such as perceived lack of time, have also led to more convenience retail, and this is particularly prevalent in the grocery sector. To manage the increasing complexity of their businesses, retailers have developed new strategies to manage the shopper experience and ensure that resources are used effectively. Originally, the implementation of so-called multichannel retailing was heralded as the way forward and a result of the growth of digital retail, and new ways of shopping proliferated. Now these new channels are being integrated with traditional shopping journeys, which has led to the advent of new omni-channel strategies (Rigby, 2011) across the retail sector. The term "omni-channel" is usually used to define the means of offering the customer a seamless purchase experience, however and wherever they choose to shop.

This change in the way that consumers shop is now requiring the grocery sector to innovate ever faster just to keep trading profitably and to keep their shoppers happy. However, this shift in purchase habits and delivery options has implications for the requirements of food and beverage products themselves. Whereas in the recent past, the product development team could assume that grocery products would be purchased primarily in a supermarket, this is no longer the case. Product selections can take place at home, in large-store formats, or in convenience outlets, to name but a few. This leads to new and different requirements for the successful product and pack. So, it is increasingly important to consider both the pack/product experience and the shopping experience for successful food and beverage sales and satisfied customers.

The design of primary, secondary, and tertiary packaging for the emerging omni-channel environment must be integrated throughout the business to ensure that the efficiency of increasingly complex logistics is optimized. New criteria to determine shelf impact and differentiation will be needed as the shopper evolves to operate across multiple channels. This will impact both the structural and functional elements of the packaging design and make it even more important that the interactions between the pack and the consumer are well understood. Such a complex design process will

Integrating the Packaging and Product Experience in Food and Beverages. http://dx.doi.org/10.1016/B978-0-08-100356-5.00004-8

mean that the role of the packaging technologist must be extended to cover all of these new requirements (Azzi et al., 2012).

This chapter concentrates on understanding the impact that this emergent omni-channel retail paradigm will have on innovation in the food and beverage industry, and how this might change the consumer experience. We will focus our discussions primarily on the impact that these new trends will have on the packaging of food and beverages. First, we will explore the world of omni-channel retailing and show how these new shopper journeys are having a significant impact upon grocery shopping. Then we will look at two areas where additional considerations must be included in the product development process to meet the requirements of the omni-channel retail environment, namely:

- How does the packaging attract the consumer at the point of sale, when that may be a retail store but could equally be a digital image?
- How can packaging withstand the rigors of new supply chains and being shipped in both large and small volumes around the globe?

The analysis of the new requirements will show how packaging is becoming a more significant part of the marketing mix because it has persistence across channels. We will show how this provides an opportunity to use the pack as a tool to integrate the customer experience across many shopping channels. The chapter will close with the presentation of a packaging road map for omni-channel retail, which includes all the different considerations necessary to successfully integrate the pack and product design and ensure suitability for the emerging food and beverage omni-channel environment.

4.2   The Omni-Channel Shopping Experience

As e-commerce grows in popularity, the number of different ways (or channels) that consumers can purchase their products is proliferating, and the shopper journey has many new routes and possibilities. In order to maximize sales across all of these channels, retailers originally developed multichannel retailing strategies to help them manage and grow their businesses. Multichannel retail is where a number of channels are provided through which consumers can purchase goods, in particular online, stores, and catalogs. However, as the number of channels increased along with the potential complexity of the shopper journey, it required retailers to rethink their strategies. This led to the advent of the omni-channel approach, which puts more emphasis on the interplay between the channels (Neslin et al., 2014). Omni-channel management was defined as "The synergetic management of the numerous available channels and customer touchpoints, in such a way that the customer experience across channels and the performance over channels is optimized" (Verhoef et al., 2015). Thus omni-channel retailing is about ensuring that the customer has the same experience of the retailer, however and whenever they choose to shop. This also requires an integrated perspective of all the communication channels between shopper, brand, and retailer.

It is claimed that by focusing on the total shopping journey, omni-channel strategies deliver a more seamless experience for consumers, leading to greater satisfaction and shopper retention (Shankar et al., 2011). Coordination of physical and digital shopping experiences has been proven to increase profits (Chatterjee, 2007; Avery et al., 2012; Oh and Teo, 2006) and hassle-free transactions in the omni-channel world increase consumer trust. Hongyoun Hahn and Kim (2009) and Cao and Li (2015) identified five mechanisms that affect sales from implementing omni-channel retailing. These are (1) improved trust, (2) increased loyalty, (3) higher consumer conversion rates, (4) greater opportunities to cross-sell, and (5) the loss of special channel features.

As the omni-channel approach involves more and more retail channels, the distinctions between the channels are becoming blurred. This is primarily because most modern shoppers do not think about the channel they are using and hop from one to another interchangeably during the search and purchase process, and thus learn how to shop differently in a more "omni-channel" way (Neslin et al., 2014). It is virtually impossible for firms to control this behavior and in fact there is evidence that they should encourage this to increase loyalty (Schramm-Klein et al., 2011) and sales (Cao and Li, 2015). By way of an example, many shoppers now spend time researching their purchase in one channel and make the decision to buy in another (eg, online and in-store or vice versa). This had led to particular shopper behaviors being defined called "Showrooming" and "Webrooming," as seen in Fig. 4.1 (Chatterjee, 2010; Zhang and Oh, 2013):

- Showrooming is where the customers visits a physical store to gather product information and simultaneously searches on a mobile device to get more information about prices, deals, and offers (customers often finally purchase online).
- Webrooming, often seen as the opposite of showrooming, where shoppers research online for information about their potential purchase but ultimately purchase in a physical store.

In reality, retailers are now offering customers defined opportunities to cross the channel boundaries using strategies to enhance omni-channel benefits. One method which has proven successful, particularly in the United Kingdom, is so-called "click and collect" which is where shoppers order online but collect goods in a store. In 2014, 35% of all UK Internet purchases were through this method, with another 39% opting for a "reserve and collect" option (Mintel, 2014). Evidence has been presented that this approach can reduce such showrooming behavior (Chatterjee, 2010), but it is also suggested that a lot more research is needed before we can fully understand how this impacts purchase decisions (Heitz-Spahn, 2013).

The latest technological advances in mobile computing are further blurring boundaries between traditional retail environments and the digital retail world (Brynjolfsson et al., 2013). Smartphones, and more recently watches, now enable retailers to become more innovative and support the cross-channel interaction with consumers through multiple touch points. For example, some retailers are using location-based services to offer customers electronic coupons on their phones the moment they enter a store (or perhaps a competitor's!), and others have free Wi-Fi in-store so that consumers can scan QR codes to see online product reviews and exclusive video content. Some manufacturers and retailers have experimented with supporting technologies that overlay (or augment) the real world with digital content such as graphics or sounds. This is called augmented reality (AR) (Montgomery, 2013; GSMA, 2014). Most AR

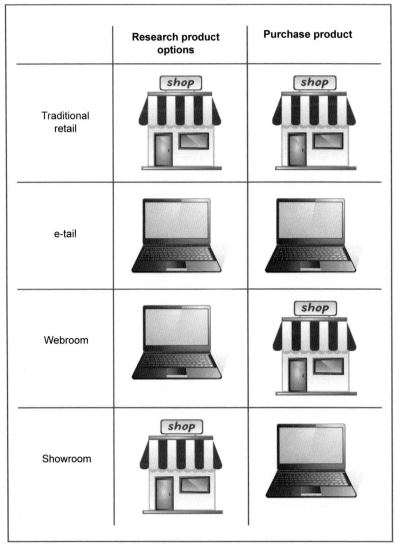

Figure 4.1 Showrooming versus Webrooming.

applications require the user to look through the camera of a smart device which shows the real world with additional images, changing the user perspective to tell a brand story (see, eg, https://www.youtube.com/watch?v=Go9rf9GmYpM and https://www.youtube.com/watch?v=z5geX7mqGAs). This technology allows retailers and manufacturers to link the digital and physical worlds and deliver a rich blend of experiences, although so far the applications of AR have not fully been accepted by customers and have still got some way to go to convince the sector that the investment is financially beneficial above a brand-building exercise. Perhaps the technology needs to mature before full business benefits are realized.

The changing face of retail and new shopping behaviors brought about by the introduction of omni-channel retailing are felt throughout the supply chain. While manufacturers will find it increasingly difficult to predict product volumes and logistics requirements for different retailers because of the complex channel dynamics, perhaps there are new opportunities. The boundaries between manufacturing and retailing are becoming less clear through direct selling and collaborations to develop exclusive products (Brynjolfsson et al., 2013).

The increasing number of shopping channels means that purchasing decisions can be shaped by information from many sources such as the store, Web sites, mobile apps, or social media. However, as Brynjolfsson et al. (2013) point out, a further influential channel is the product itself and the way that it is packaged in such a way as to deliver product knowledge and integration with other channels. This is an area that, as yet, is little exploited by manufacturers and offers a new route to business growth. Thus, it is imperative that, as the retail sector as a whole is moving from a multichannel to omni-channel view, food and beverage manufacturers also keep pace with this transition and better understand how the new shopping behaviors and supply chain requirements could influence their product and packaging development and identify the opportunities for innovation.

4.2.1   Omni-Channel Grocery Retail

We have seen that research in the area of omni-channel retailing is still very much in the early stages and there are many open questions, particularly when trying to understand and predict shopper behavior and purchase decisions. However, what is even less well understood is how the new omni-channel management strategies being implemented at the retailer level will impact food and beverage manufacturers within their supply chains. These manufacturers must understand how they can use knowledge of the complex omni-channel environment to maximize their business. This section looks at the current trends in grocery retailing, particularly food and beverage products, and unpicks the key opportunities for future innovation.

Online shopping for groceries is somewhat lagging behind other retail categories (Melis et al., 2015). Online shopping accounts for 4.1% of the overall grocery market in the United Kingdom and is predicted to rise to 6.5% by 2018. Tesco has the largest online market share (40.3%), followed by Sainsbury's (16.7%) and Asda (13.9%) (Mintel, 2015). This split is unsurprising, as research has shown that customers tend to choose the online channel of their preferred retailer first. However, when online shopping experience increases, shoppers start to switch between the online options of the different retailers (Melis et al., 2015). While current and predicted market share are not yet significant, almost 50% of us claim to do at least some of our food and beverage shopping online (Mintel, 2015). This could be accounted for, in some measure, by the rise in online specialist food sites such as Graze (https://www.graze.com/uk) (on the go snacking) and Hello Fresh (https://www.hellofresh.co.uk/) (scratch cooking meals in a box). This means that food and beverage manufacturers ignore the digital sales of their products at their peril.

Traditionally, retailers defined their business with three simple imperatives: stock products you think your target customers will want, cultivate awareness of what's in the store, and make it enticing and easy for them to buy (Rigby, 2011). In contrast, successful retailing in an omni-channel world is a much more complex business. To thrive in today's markets, retailers must innovate to devise different ways of attracting their target customers. Some shoppers are still happy to be served much the way they were in the past (Brynjolfsson et al., 2013). However, most consumers expect to be able to buy products 24/7 in a location of their choice, and food and beverage shopping is no exception. The large out-of-town superstore, where families made the weekly trip to purchase all our necessary items, is becoming a thing of the past. Our increasingly busy lifestyles mean that smaller, more frequent food shopping trips (often called "top-up shops") are becoming the norm and online shopping is growing in popularity. In addition to this, there are now also an increasing number of options for the customer to collect their grocery shopping once ordered online. Drive through, click and collect, self-service lockers, home delivery, and local/convenience stores are all growing options and are further contributing to the decline of the large out-of-town superstore and the rise in omni-channel grocery shopping habits driven by new technologies and convenience.

Notwithstanding this, there are a number of barriers to the growth of online retailing in grocery. Today's shopper does not like paying for the delivery of their food shopping, and the economic situation is that the business model of online sales for UK supermarkets does not currently make financial sense. That is, it costs more to deliver the shopping than businesses are able to charge customers. This is compounded by the fact that customers tend to keep to their shopping lists and make few impulse purchases, decreasing the average spend per head (Mintel, 2015). Thus, irrespective of customer demand, it is not currently in the interests of retailers to increase online provision until the business case can be improved (KPMG/IPSOS Retail Think Tank, 2014).

The second barrier is the issue of consumer trust. Grocery shopping is inherently an emotional and sensory experience. Consumers like to sniff, touch, and see the food and beverage products they are purchasing to sustain and look after their family. Outsourcing the selection of food (particularly fresh produce) to a supermarket picker is quite a large step for many shoppers. In addition, there are concerns about the length of sell-by dates and idiosyncratic product substitutions mean that many shoppers are not willing to shop online for their groceries.

Globally, the situation is a little different. Continued increases in mobile adoption and broadband penetration, particularly in developing regions, have helped boost online grocery sales (Nielsen, 2015). More than one-third (37%) of Asia-Pacific respondents, and even more in China (46%), say they use an online ordering and delivery service. In China between 2013 and 2014, e-commerce sales increased by 40%, and food was the primary growth sector. The reasons for this growth are clear. The region's rapid urbanization and high population density make the home delivery model economically viable, particularly when coupled with low labor costs.

While the low current market share for online groceries, particularly in the United Kingdom, suggests a long-term slow-growth strategy, customers are rapidly learning how to shop in an omni-channel way, and behaviors and expectations are changing. Our younger generations have grown up with online retail and have fewer trust issues.

As they mature, move into their own households, and become responsible for their own grocery shopping, there will be much greater demand for online groceries, and the retailers will have to respond.

4.3   Innovative Packaging for Omni-Channel Retail

Packaging has a really important role to play in a traditional retail environment. It must be designed to (a) protect and preserve the product throughout the rigors and challenges of the packaging supply chain (see Fig. 4.2) and (b) effectively market the product at the point of sale (Prendergast and Pitt, 1996). This is in addition to environmental concerns such as reuse or recycling and usability requirements such as resealability and ease of opening.

In the grocery sector in particular, a great deal of time and resources are expended at both the retailer and the manufacturer to ensure that all food and beverage packaging meets these exacting requirements. And companies have gotten really good at developing packaging that can deliver to meet these current needs. However, the opportunities offered through the new omni-channel retailing strategies mean that the current design and development heuristics for packaging now require further consideration. Food and beverage manufacturers who can understand (and perhaps predict) future consumer behavior have an opportunity to realize real innovation and develop packaging that better suits the needs of the omni-channel shopper. This brings its own challenges, and the following sections will look at this in more detail, namely:

- How does the packaging attract the consumer at the point of sale, when that may be a retail store but could equally be a digital image?
- How can packaging withstand the rigors of new supply chains and being shipped in both large and small volumes around the globe?

4.3.1   Packaging for Omni-Channel Marketing

In today's complex omni-channel retail environment, it is critically important that the marketing role of a pack is as equally effective when viewed online as it is in the store. Although, this is not as easy as it sounds. This section focuses on this issue and discusses the influence of different shopping channels on how consumers select products and what this means for the packaging itself.

Online grocers offer a number of different ways for consumers to find products on their site. A search term can be entered (eg, "baked beans"), the shopper can browse through a list of special offers, or they can navigate through the tabs/links to find a

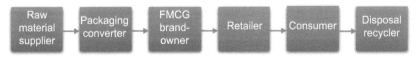

Figure 4.2 The packaging supply chain.

particular virtual department or aisle. Research has shown that 95% of consumers use the navigation route, 80% use the search facility, and 66% browse the special offer pages (Benn et al., 2015).

The navigation route has two stages. First, the consumer needs to move through a hierarchy of departments until they arrive at the specific virtual aisle, before selecting the particular product to place in their basket. This is usually done by sorting by an attribute such as price or by manually scrolling and scanning down the (often quite long) list. This means that the digital image of the product is critically important for communicating information to the consumer to allow them to make the purchase decision. The product image is typically a small 2-D thumbnail, usually of the packaging with some limited explanatory text to support the consumer's decision. Customers can click through for further information about the product, but it has also been shown that consumers tend to only use the image of the product to make their decision rather than reading the detailed information (Bamburry, 2015; Benn et al., 2015). It is clear that accessing the information could support the purchase decision; however, it is mostly the clarity of the text font that enhances the experience (Mosteller et al., 2014). This demonstrates that the clarity of the image and supporting text have great importance in influencing the consumer's purchase decision.

The grocery product image that is presented on a retailer's Web site is usually of the packaging rather than the product itself. The obvious benefit of this is that the image closely replicates the in-store product view that consumer has prior experience of. This increases the physical and mental tangibility of the product which in turn reduces the perceived risk of online purchasing for consumers (Nepomuceno et al., 2014). Designers have been very successful at harnessing complex psychological techniques, developed over many years, to produce packs that stand out on a physical shelf and encourage consumers to pick up the product and ultimately purchase in-store. We know that the visual pack design has significant impact upon perceived quality and brand preference (Wang, 2013), hence the amount of time and resource spent by businesses getting this right. However, the rules for online selection and purchase may be different and the same techniques might not necessarily apply in this situation. Fig. 4.3 illustrates this point showing the visual difference between product selection online and in-store.

(A)   (B)

Figure 4.3 The visual difference between (A) in-store and (B) online product selection.

While imaging the pack online clearly aids navigation for familiar products, this misses innovative opportunities to take advantage of the functions of the digital environment. For example, research has shown that providing an interactive 3-D image increases users' enjoyment and trust in the product (Algharabat and Dennis, 2010). It has also been shown that multiple images can be successfully used to improve sales through an online retail store and that the larger the image, the easier it is for the consumer to select and purchase a product (Song and Kim, 2012).

Increasing the quality and number of images may be tricky to implement on a grocery Web site with a large number of products and variants to show. However, there are obvious areas that could be improved. For example, some packaging includes photographs of the product within the container to show consumers what to expect. This is particularly prevalent within the ready meal category, see Fig. 4.4. Currently, the Web site thumbnail is a photographic image of the pack showing a photographic image of the product. Perhaps a simple solution could be showing the pack and the product image directly adjacent, which might offer a better shopping experience. Whether this is workable or not, the key message is that retailers and manufacturers need to think about how to best represent the product online and not just default to the "take a photograph of the pack" strategy.

Because it is so small, the online product image is often found to be difficult to interpret, especially for pack size and in identifying copycat brands. This is a particular problem on the smaller screens of mobile devices, because as customers transfer to more habitual mobile shopping, their value to the grocer increases (Wang et al., 2015). There are many anecdotal examples of frustrated customers receiving huge packs when expecting small ones and vice versa because the different variants have the same graphics, colors, and the outline or profile. The difficulty of selecting the correct size and variant of a product is shown in Fig. 4.5. While the strategy of maintaining a design theme for a brand works well for navigation and product selection in the physical store, when designing packaging for online purchase, perhaps businesses need to employ different rules. Using color to communicate different pack variants might be a more effective online strategy, and perhaps this could also support the physical navigation on the ever more cluttered supermarket shelves. Another route could be to offer

(A)   (B)

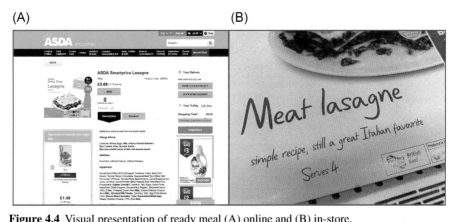

Figure 4.4 Visual presentation of ready meal (A) online and (B) in-store.

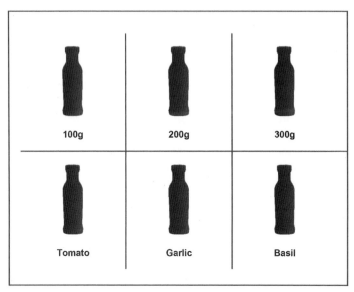

Figure 4.5 The difficulty of identifying product variants with the same silhouette online.

different pack structures for product variants to create an easy-to-identify silhouette. This would mean that designers need to consider both the 3-D pack structure and a relevant 2-D outline for Web site navigation. Furthermore, this will require food and beverage manufacturers to include the 2-D online representation within their brief to packaging designers.

In-store, the consumer can use many different communication channels to explore the product. This is where the sensory experience becomes really important in supporting the purchase decision. To select a product to buy in-store, consumers can see, touch, smell, hear, and even occasionally taste both the product and the pack. However, online this sensory communication is reduced to just visual and relies upon a consumer's prior experience to make assumptions about the other four sensory attributes. We have seen in the previous discussion how important it is to get the image right, but how can packaging designers engage the other senses?

When a consumer chooses to do their grocery shop online, other than any previous purchases, the first physical (and thus true multisensory) experience they have of the products they have bought is when they are delivered to their door or they pick up their groceries from a click and collect facility. This first encounter can impact any future purchase and raises the importance of this consumer touch point in the packaging value chain. Thus, the point of fulfillment (and subsequent product use) now becomes the key opportunity for sensory delight in omni-channel retail, influencing the repurchase decision.

Packaging provides brand reassurance at the point of fulfillment, and if the groceries are shipped in secondary packaging, this can be used to add further value through additional marketing/other information or free samples. However, probably the first experience that a consumer has of interacting with the primary pack is when they pick it up and remove it from the secondary pack or the carrier bag, elevating the tactile appeal of the primary pack surface to become one of most significant sensory experiences (Chen et al., 2009). The consideration of this and other sensory communications must be included within the packaging engineer's brief and design scope. This will ensure that the fulfillment experience is congruent with the brand message and delights the consumer, creating a memorable encounter that will support any repurchase decision.

4.3.2 Packaging for Omni-Channel Transport

There are many new requirements placed on manufacturers and retailers by omni-channel strategies which significantly change how products are moved around the world and delivered to consumers (Knott, 2015). This clearly will have implications for the primary, secondary, and tertiary package requirements.

No longer do customers only shop in-store and take home themselves; there are many (and increasing) delivery and fulfillment options to suit customer needs. The food and beverage sector is no different from any other retail sector and is having to implement new ways for customers to access its products. In Europe, these new fulfillment options include buy online/deliver to home, buy online/pick up in store, buy online/collect via locker, and many more variants of these (Samuel, 2012).

With the online grocery market in the United Kingdom expected to be worth £13.8 billion by 2019 (Mintel, 2015), it is really important that grocery retailers are able to offer this range of options to their customers. However, it is inevitable that these new order fulfillment strategies, illustrated in Fig. 4.6, will lead to challenges and innovation opportunities for the manufacturers who are providing the products that travel through these new supply chains.

Most primary packaging is designed to be efficiently transported around the world to the retail stores in transit boxes, piled high on pallets, and usually shrink-wrapped. Primary packaging is optimally designed from a strength and volume perspective to be suitable for this system. In a traditional grocery supply chain, where products leave a manufacturer to go to a national distribution center (NDC) and then sent to individual stores in large lorries, this fulfills the requirements of protection usually very well, and most products reach the store shelves intact. Fig. 4.7 illustrates a standard grocery supply chain.

In omni-channel retailing for the grocery sector, this is very different; there is an additional need to transport the food and beverages from a store or a distribution warehouse to the customer's home (or a click and collect facility/locker). In most cases, prior to this final delivery, the customer's order is picked and placed in reusable boxes or carrier bags, and no other discernable transit packaging is used (Fig. 4.8 shows a typical example of this). This final leg of the product journey is often termed "last-mile logistics" (Gevaers et al., 2008) and is very different to the traditional retailing model whereby the customer takes responsibility for ensuring that the product reaches their home without mishap.

This is a critical step in the customer journey and essential for a great omni-channel experience, but currently is not always cost-effective (Gevaers et al., 2014) and offers

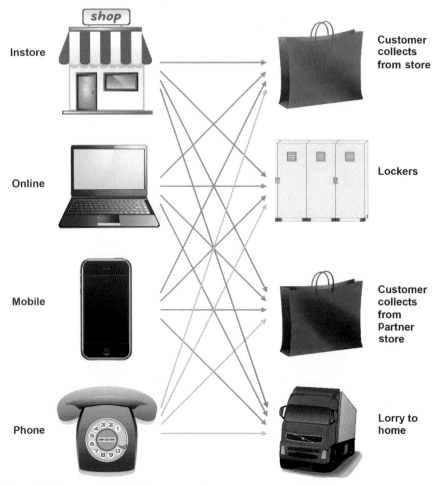

Figure 4.6 Customer order fulfillment options.

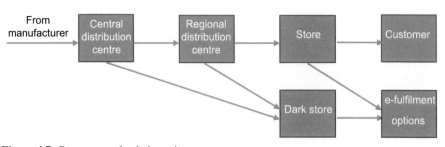

Figure 4.7 Grocery supply chain options.

a number of key challenges (Boyer et al., 2009). Both manufacturers and retailers must ensure that their packs are suitable for this final step in the supply chain and allow the product to reach the customer intact. As we have seen in the previous section, again, this places new requirements on the packaging design process.

Figure 4.8 Groceries for home delivery.

Food and beverage products that need to be transported through omni-channel retailing supply chains must be able to withstand the rigors of the system. First, the pack must have enough structural integrity to be transported individually and not break without suitable secondary or transit packs, even when in carrier bags with other products. The packaged products must be suitable for placing in collection lockers, widely in use around the United Kingdom and elsewhere. Walmart is also looking to its UK operation, Asda, to lead innovation in this area and is trialing "click and collect pods," a fully automated click-and-collect facility operated by robots (Chaudhuri, 2015).

As the omni-channel system matures, some manufacturers may also consider direct selling to their consumers, and so the packaging must also withstand being posted and be suitable for any returns for faulty or unwanted goods. This requirement may not be too far away. Amazon, the large online retailer are already gearing up to do this in the United States with PrimePantry http://www.amazon.com/gp/pantry/info – "a unique shopping experience on Amazon.com. Prime members can shop popular household essentials and have them conveniently delivered." and AmazonFresh (https://fresh.amazon.com/welcome) "Free same-day and early morning delivery of orders over $35 on thousands of items, including fresh grocery and local products." There are some industry reports that this will come to the United Kingdom in late 2015 (Hobbs, 2015).

Because there are more routes to the customers through, for example, NDCs, convenience stores, lockers etc., typically smaller volumes are being delivered at any one time (Smillie, 2015). No longer can it be assumed that the product will be moved in a well-stacked pallet in a large lorry. Pallets will get broken down earlier in the supply chain and more mixed loads of packaged goods will be shipped around, often in smaller vans. This calls into question assumptions about suitable shipping volumes and sizes, and perhaps even new ways of palletization and transit packs may be needed.

The increasing complexity of the omni-channel supply chains make it potentially more difficult to control the environment, which is a particular problem for fresh produce such as food and beverages. Perhaps more packages will need to include time temperature indicators or other similar sensing and monitoring devices to ensure that the consumer has confidence of the freshness of the products on receipt. In extreme cases, perhaps packaging technologists might need to consider including environmental controls within the pack, such as additional insulation. There are also other issues which arise from the additional supply chain routes. Either primary or secondary packaging may need to be designed with supplementary security and tracking devices to reduce losses, guard against counterfeit goods, and retain consumer confidence in the product that reaches them. These added features are clearly not without additional cost implications and will have to offer significant benefits for manufacturers to implement widely.

However, regardless of any supply chain and logistics issues encountered, the packaging technologist must, as always, deliver their innovations within tight cost constraints, being careful to minimize the number of pack variants that the complex retail sector might suggest. In addition, and perhaps most importantly, they need to achieve all these new requirements without adding significantly to the amount of packaging in use. More packaging only adds to the environmental burden and cost for the manufacturers and consumers will not accept what they perceive as unnecessary packaging.

However, the world will not stand still, and more new ideas will be implemented as the omni-channel retail model becomes more accepted and technology advances at a rapid pace. There are more possible changes on the horizon. Amazon has been widely reported as trialing delivery drones to get products to customers' doors (Bamburry, 2015). We can only speculate what the packaging implications may be for these and other future innovations!

4.4   Packaging as the Omni-Channel Integrator

One of the key attributes of an omni-world is that all retail channels are integrated and deliver a seamless experience to the customer. It is not fully clear how this can be achieved and research into the many different aspects is still in the very early stages (Verhoef et al., 2015). The delivery of a seamless experience to consumers is not trivial and depends very much on the retailer putting the customer at the heart of everything they do. But as shoppers are now happy to combine browsing online with popping into a physical store, what are the common denominators that can support retailers to provide this seamless experience?

Within the grocery sector, one of the factors that is common to all the channels is the product and its packaging. By focusing on this commonality, perhaps strategies could be developed to enhance the channel integration. It would be difficult to change products, particularly food and beverages; however, there is a great deal of flexibility to design the packaging to be able to cross the channel boundaries.

In Section 4.2, AR was discussed as a technology that overlays a view of the real world with digital content such as graphics or sounds, thus creating an experience that links the digital and physical worlds. AR allows retailers to respond in a positive way to "showrooming," by linking images printed on the packaging of products to customers' smartphones or devices (Olsson et al., 2013). Many food and beverage brands have utilized this technology to engage with consumers (see https://blippar. com/en/blipp for some examples). However, a more recent AR development includes tactile technology that allows for change to the tactile feeling of real objects by augmenting them with virtual tactile textures using a device worn by the user (Bau and Poupyrev, 2012).

Bar codes and QR codes have been widely used to allow smart devices to scan products in the home to create shopping lists, in-store to identify products in the basket for purchase or other similar ideas (Bornemann, 2012). Amazon have taken this a step further and their app uses image recognition technology to identify the silhouette of a product to create shopping lists (McCracken, 2014). While this is still in the early stages of development, it shows just how important the pack is to link the real and digital world and how these requirements should all be considered within the development process.

Finally, Ericsson have recently announced a new technology called Connected Paper (Dyreklev, 2014). It uses "capacitive coupling" technology, which transfers an electrical signal through the body, using a similar principle as a phone's touchscreen, responding to the finger's proximity rather than the physical pressure exerted. By simply touching a piece of paper (or packaging), relevant information is instantly displayed on a smart device. There is no need to tap the pack with the device (like using near field communication), it happens without intervention.

4.5 Satisfying Customers Through Omni-Channel Packaging Innovation

Throughout this chapter, we have shown how the retail world is changing to become a mix of digital and physical shopping and that consumers are expecting a seamless experience, whatever way they choose to purchase goods. Online purchase and delivery of groceries is still in its early stage, but is widely predicted to grow significantly over the next 5 years. The previous sections have identified many of the key challenges and opportunities that this brings.

We have shown how packaging has become a more significant part of the marketing mix that is persistent across the channels. This provides an opportunity to see the pack as a tool to integrate the customer experience across the many shopping channels. Through good design and considered use of new technologies, the pack can play a key role in enhancing the customer experience. This will not be easy to do, and many of the current rules for packaging design need to be reassessed. Packaging technologists must be aware of all these new requirements to ensure that they create packaging that is both fit for purpose and enhances the customer satisfaction at all points in their retail experience.

To summarize all the key points discussed in this chapter, a packaging road map for omni-channel retail is presented below. This road map includes all the different considerations necessary to successfully integrate the pack and product design and ensure suitability for the emerging food and beverage omni-channel environment.

- Place the same emphasis on the products' digital presence (image and associated text) as on the packaging structure, copy, and graphics. Ensure the image(s) is of good quality and consider how it could be more three-dimensional and/or interactive.
- Design a unique pack silhouette (and consider how it could be changed for pack size and product variant) to aid the online navigation and product selection of your consumer.
- Pay particular attention to the tactile feel of the pack to enhance the delivery experience.
- Make sure the pack has sufficient structural integrity to be shipped individually and withstand the rigors of transportation in a carrier bag.
- Understand the detailed characteristics of the supply chain environment, right to your consumers' door, and make sure the packaging protects the product in all circumstances. If there is any doubt, consider using a sensing technology to reassure the consumer of the freshness of your product.
- Map the different order fulfillment options of the retailers and ensure your pack is suitable to be shipped in every situation.
- Consider using new technologies, particularly for mobile shoppers, to increase brand engagement and help the pack to create a seamless customer experience.
- And finally—make sure that, wherever possible, none of the changes and innovations increase the overall amount of packaging needed.

4.6 Summary

In the new world of omni-channel retailing, there many opportunities for food and beverage manufacturers to create innovative packaging that is better able to deal with the needs and challenges of this emergent shopper behavior. While online shopping only accounts for 4.1% of the overall grocery market in the United Kingdom, it is widely predicted to rise and should be ignored by food and beverage manufacturers at their peril.

This chapter has shown how important it is that the remit of the packaging technologist is expanded to ensure that the packs developed are fit for the new retail environment. We have seen how critical the image used to select products on a retailer's Web site is and how the point of delivery to the customer has increased in importance. The implications for the pack when being transported through the more complex supply chains has been investigated as well as how important it is to understand the new environment, particularly for food and beverage products. We have presented some ideas on the future role of packaging as a channel integration tool using emerging smart technologies. To conclude the chapter, a set of design requirements has been developed that can be used to inform future packaging design for an omni-channel retail world.

References

Algharabat, R., Dennis, C., 2010. Using authentic 3D product visualisation for an electrical online retailer. Journal of Customer Behaviour 9 (2), 97–115.
Avery, J., Steenburgh, T.J., Deighton, J., Caravella, M., 2012. Adding bricks to clicks: predicting the patterns of cross-channel elasticities over time. Journal of Marketing 76 (3), 96–111.

Azzi, A., Battini, D., Persona, A., Sgarbossa, F., 2012. Packaging design: general framework and research agenda. Packaging Technology and Science 25 (8), 435–456.

Bamburry, D., 2015. Drones: designed for product delivery. Design Management Review 26 (1), 40–48.

Bau, O., Poupyrev, I., 2012. REVEL: tactile feedback technology for augmented reality. ACM Transactions on Graphics 31 (4), 1–11.

Benn, Y., Webb, T.L., Chang, B.P., Reidy, J., 2015. What information do consumers consider, and how do they look for it, when shopping for groceries online? Appetite 89, 265–273.

Bornemann, E., 2012. Best QR code innovations. Information Today 29 (2), 10.

Boyer, K.K., Prud'homme, A.M., Chung, W., 2009. The last mile challenge: evaluating the effects of customer density and delivery window patterns. Journal of Business Logistics 30 (1), 185–201.

Brynjolfsson, E., Hu, Y.J., Rahman, M.S., 2013. Competing in the age of omnichannel retailing. MIT Sloan Management Review 54 (4), 23–29.

Cao, L., Li, L., 2015. The impact of cross-channel integration on retailers' sales growth. Journal of Retailing 91 (2), 198–216.

Chatterjee, P., 2007. Cross-channel product ordering and payment policies in multichannel retailing: implications for shopping behavior and retailer profitability. Journal of Shopping Center Research 13 (2), 31–56.

Chatterjee, P., 2010. Causes and consequences of 'order online pick up in-store'shopping behavior. The International Review of Retail, Distribution and Consumer Research 20 (4), 431–448.

Chaudhuri, S., 2015. Wal-Mart Looks to U.K. Unit for Lessons on Online Groceries. Wall Street Journal, p. 1 (Online).

Chen, X., Barnes, C., Childs, T., Henson, B., Shao, F., 2009. Materials' tactile testing and characterisation for consumer products' affective packaging design. Materials & Design 30 (10), 4299–4310.

Dyreklev, P., 2014. Printed Electronics Meet Mobile. Retrieved 10/9/15, from: https://www.acreo.se/projects/printed-electronics-meet-mobile.

Gevaers, R., Van de Voorde, E., Vanelslander, T., 2008. Technical and process innovations in logistics: opportunities, barriers and best practices. In: European Transport Conference 2008; Proceedings.

Gevaers, R., Van de Voorde, E., Vanelslander, T., 2014. Cost modelling and simulation of last-mile characteristics in an innovative B2C supply chain environment with implications on urban areas and cities. Procedia-Social and Behavioral Sciences 125, 398–411.

GSMA, 2014. Tesco Testing IBM's Augmented Reality Mobile App for Product Placement. Enterprise Innovation.

Heitz-Spahn, S., 2013. Cross-channel free-riding consumer behavior in a multichannel environment: an investigation of shopping motives, sociodemographics and product categories. Journal of Retailing and Consumer Services 20 (6), 570–578.

Hobbs, T., 2015. Amazon Fresh: Should the UK Supermarkets Be Afraid? Marketing Week, p. 1 (Online Edition).

Hongyoun Hahn, K., Kim, J., 2009. The effect of offline brand trust and perceived internet confidence on online shopping intention in the integrated multi-channel context. International Journal of Retail & Distribution Management 37 (2), 126–141.

KPMG/IPSOS Retail Think Tank, 2014. The Future of the Grocery Sector in the UK. Retrieved 10/9/15, from: http://www.kpmg.com/uk/en/issuesandinsights/articlespublications/news-releases/pages/the-future-of-the-grocery-sector-in-the-uk.aspx.

Knott, M., 2015. Meeting the Multi-Channel Challenge. Food Manufacture. William Reed Business Media Ltd, p. 39.

McCracken, H., 2014. Amazon's iPhone App Uses Image Recognition to "See" Real-World Products You Want to Buy. Time.com, p. 1.

Melis, K., Campo, K., Breugelmans, E., Lamey, L., 2015. The impact of the multi-channel retail mix on online store choice: does online experience matter? Journal of Retailing 91 (2), 272–288.

Mintel, 2014. Click-and-Collect – UK.

Mintel, 2015. Online Grocery Retailing – UK.

Montgomery, A., 2013. 'Exploding' Fruit Drink Packaging from Williams Murray Hamm. Design Week, p. 7 (Online Edition).

Mosteller, J., Donthu, N., Eroglu, S., 2014. The fluent online shopping experience. Journal of Business Research 67 (11), 2486–2493.

Nepomuceno, M.V., Laroche, M., Richard, M.-O., 2014. How to reduce perceived risk when buying online: the interactions between intangibility, product knowledge, brand familiarity, privacy and security concerns. Journal of Retailing and Consumer Services 21 (4), 619–629.

Neslin, S., Jerath, K., Bodapati, A., Bradlow, E., Deighton, J., Gensler, S., Lee, L., Montaguti, E., Telang, R., Venkatesan, R., Verhoef, P., Zhang, Z.J., 2014. The interrelationships between brand and channel choice. Marketing Letters 25 (3), 319–330.

Nielsen, 2015. The Future of Grocery. Retrieved 10/9/15, from: http://www.nielsen.com/cn/en/insights/reports/2015/the-future-of-grocery.html.

Oh, L.-B., Teo, H.-H., 2006. A Value-Based Approach to Developing a Multi-Channel Shopper Typology. ECIS.

Olsson, T., Lagerstam, E., KärkkäinenK, T., Väänänen-Vainio-Mattila, K., 2013. Expected user experience of mobile augmented reality services: a user study in the context of shopping centres. Personal Ubiquitous Computing 17 (2), 287–304.

Prendergast, G., Pitt, L., 1996. Packaging, marketing, logistics and the environment: are there trade-offs? International Journal of Physical Distribution & Logistics Management 26 (6), 60–72.

Rigby, D., 2011. The future of shopping. Harvard Business Review 89 (12), 65–76.

Samuel, S., 2012. Building Supply Chain Capability for a Multi-Channel Future. Retrieved 10/9/15, from: http://www.igd.com/Research/Supply-chain/Logistics/4959/Building-supply-chain-capability-for-a-multi-channel-future/.

Schramm-Klein, H., Wagner, G., Steinmann, S., Morschett, D., 2011. Cross-channel integration—is it valued by customers? The International Review of Retail, Distribution and Consumer Research 21 (5), 501–511.

Shankar, V., Inman, J.J., Mantrala, M., Kelley, E., Rizley, R., 2011. Innovations in shopper marketing: current insights and future research issues. Journal of Retailing 87, S29–S42.

Smillie, D., 2015. Forecasting Demand in a World of Uncertainty. Retrieved 10/9/2015, from: http://www.igd.com/Research/Supply-chain/Strategy-planning-technology/29295/Forecasting-demand-in-a-world-of-uncertainty/.

Song, S.S., Kim, M., 2012. Does More Mean Better? An Examination of Visual Product Presentation in E-retailing.

Verhoef, P.C., Kannan, P.K., Inman, J.J., 2015. From multi-channel retailing to omni-channel retailing: introduction to the special issue on multi-channel retailing. Journal of Retailing 91 (2), 174–181.

Wang, E.S.T., 2013. The influence of visual packaging design on perceived food product quality, value, and brand preference. International Journal of Retail & Distribution Management 41 (10), 805–816.

Wang, R.J.-H., Malthouse, E.C., Krishnamurthi, L., 2015. On the go: how mobile shopping affects customer purchase behavior. Journal of Retailing 91 (2), 217–234.

Zhang, L., Oh, L.-B., 2013. Determinants of multichannel consumer switching behavior: a comparative analysis of search and experience products. In: WHICEB 2013 Proceedings, pp. 205–212.

Emotion Measurements and Application to Product and Packaging Development

S. Spinelli
SemioSensory Research & Consulting, Prato, Italy

M. Niedziela
HCD Research, Flemington, NJ, United States

5.1 Introduction

The collection of emotional responses in sensory and consumer studies has become more and more frequent in the last 10 years, as testified by the increasing number of both scientific papers and approaches developed in private companies to measure emotions. Emotional profiling has added a new dimension to traditional sensory profiling, and many companies have been routinely measuring emotions for years in addition to liking and sensory profiling.

One of the reasons for this success is that it is widely recognized that the measuring liking alone may fail to predict product performance in the market (Thomson, 2007; King et al., 2010; Thomson and Crocker, 2015). Gaining insight into the emotional product experience can provide additional useful information: first, the perceptive-hedonic experience of tasting routinely measured throughout liking can be investigated deeper, collecting emotional responses that color the consumer experience. In addition, several findings have highlighted the role of emotions in the decision-making process, focusing on the mix between unconscious and conscious aspects of the phenomenon (Damasio, 1994, 2003; LeDoux, 1992, 1995, 1996, 2008, 2011; Rolls, 2014; see also Köster, 2009 for a discussion of that issue in relation to food choice). Third, a key issue in product experience is the consonance between the emotions and the meanings conveyed by a product and the meanings and emotions conveyed by branding through the packaging, communication mix, and marketing strategies (Thomson, 2007).

This chapter begins with the controversial definition of what an emotion is, considering both psychological and neuroscientific perspectives, and with presentations of the main methods developed or applied to investigating and measuring emotions evoked by food and drink products in sensory and consumer studies and applied consumer neuroscience. The second part of the chapter explores the emotions-based processes of product/packaging development and optimization, discussing the impacts of emotions on expectations in consumer experience of products throughout a case study. Last, suggested topics for future research are discussed.

Integrating the Packaging and Product Experience in Food and Beverages. http://dx.doi.org/10.1016/B978-0-08-100356-5.00005-X

5.2 Emotion Measurement Methods in Sensory and Consumer Studies and Applied Consumer Neuroscience

5.2.1 What Are Emotions and What Emotions Are We Measuring?

An issue that needs to be mentioned is that at present there is no agreement on a scientific definition of what an emotion is (Frijda, 2008; Scherer, 2005; Frijda and Scherer, 2009; Mulligan and Scherer, 2012). Without consensual conceptualization of exactly what phenomenon has to be studied, debates have been proliferated without any final point. However, although there is no consensual definition of emotion, there is now rather widespread acceptance that emotions have multiple components, such as a physiological arousal, motivation, expressive motor behavior, action tendencies, and subjective feeling (Scherer, 2005).

Emotions are characterized by a response synchronization (they prepare appropriate responses to an event that disrupts the flow of behavior), rapidity of change (they continuously readjust to changing circumstances or evaluations), behavioral impact (they prepare adaptive action tendencies), high intensity, and relatively short duration. For these reasons, emotions can be distinguished by other affective phenomena such as preferences (relatively stable evaluative judgments in the sense of liking and disliking), attitudes (relatively enduring beliefs and predispositions toward specific objects or persons), moods (diffused affect states characterized by a relative enduring predominance of certain types of subjective feelings, generally characterized by low intensity and long duration), affect dispositions (stable personality traits and behavior tendencies characterized by a strong affective core), and interpersonal stances (affective styles employed in the interaction with a person or a group of persons).

Given the complex nature of emotion, it is clear that there is no golden standard method to measure emotions and that each methodology emphasizes a specific part of the phenomenon. In their wide review of the methods developed to measure emotional states in an experimental setting, Mauss and Robinson (2009) emphasized that each method is sensitive to specific aspects and best captured some but not all the aspects of the emotional states. Methodologies to study emotions are various, as much as the theories that proposed a definition of the phenomenon.

In addition, when emotions are measured with the interest of investigating consumer experience and product performance, it is apparent that the object of study has become even broader and less definite. Usually questionnaires include terms that do not indicate strictly emotions, according to any definition of the phenomena, but also moods and other affective phenomena- or feeling-related concepts. Thomson et al. (2010) introduced the term "conceptualization" to mean a conceptual association, retained in the memory and thus more durably associated with the product, and less dependent on the transient nature that characterizes emotion. This choice allows Thomson and coauthors to enlarge the conceptual profile to include terms such as "conservative," "sensual," and "youthful" (Thomson, 2015) that are not at all emotions but that are certainly of interest in characterizing the emotional experience of the products.

In the last decade, there has been a growing number of studies published in scientific journals about emotions in a sensory and consumer context. The interest is passing from the development of questionnaires and the comparison between methodologies, to more specific questions including the relationship with acceptability and sensory properties, the role of context, the comparisons between blind and branded emotional profiles, and the combination of different methodologies taken from neuroscience with more traditional approaches. See Table 5.1 for an overview.

Given the complexity of the phenomenon, the need for interdisciplinary approaches has been stressed (Köster, 2009), and along with the traditional explicit measurement developed in sensory and consumer studies, implicit measurements and neuroscientific methods are becoming more and more present. This interest is associated to recent findings of the

Table 5.1 Current Research Interests in Measuring the Emotional Profile of Food Products

	References
Relationships between sensory properties and emotions	Thomson et al. (2010), Manzocco et al. (2013), Ng et al. (2013b), Spinelli et al. (2014a), Bhumiratana et al. (2014), Kuesten et al. (2014), Jager et al. (2014), Collinsworth et al. (2014), Chaya et al. (2015)
Relationships between liking and emotions	Zeinstra et al. (2009)*, King and Meiselman (2010), Cardello et al. (2012), De Wijk et al. (2012)*, Ng et al. (2013a), Ng et al. (2013b), Bhumiratana et al. (2014), Kuesten et al. (2014), Danner et al. (2014a,b), De Wijk et al. (2014)*, Spinelli et al. (2014a), Spinelli et al. (2015), Mojet et al. (2015)*, Chaya et al. (2015), Leitch et al. (2015)*
Relationships between emotions, liking and food choice	Gutjar et al. (2014), Dalenberg et al. (2014), Gutjar et al. (2015)
Situational contexts and emotions	Piqueras-Fiszman and Jaeger (2014a,b,c, 2015), Porcherot et al. (2015)
Changes in the emotional profile in blind and branded condition	Ng et al. (2013b), Spinelli et al. (2015), Thomson and Crocker (2015), Gutjar et al. (2015)
Packaging or brand emotional responses (without tasting)	Liao et al. (2015), De Pelsmaeker et al. (2013)
Emotions at different stages of product experience	Schifferstein et al. (2013), Labbe et al. (2015)
Definition of an emotion lexicon	King et al. (2010), Ferrarini et al. (2010), Thomson et al. (2010), Ng et al. (2013a), Spinelli et al. (2014a), Thomson and Crocker (2013), Gmuer et al. (2015), Chaya et al. (2015), Nestrud et al. (2016), van Zyl and Meiselman (2015)
Methodological aspects of questionnaire and test design	King et al. (2013), Jaeger et al. (2013), Thomson and Crocker (2014), Spinelli et al. (2014a), Ng et al. (2013a)

*Indicates a study that utilizes methods others than self-report questionnaires.

last decades that have highlighted the role of unarticulated/unconscious motives and associations in consumer behavior, undermining the idea that much decision-making occurs at a rational and conscious level (Kahneman, 2003) and suggesting that emotions play a role (Damasio, 1994, 2003; LeDoux, 1992, 1995, 1996, 2008, 2011; Rolls, 2014).

Neuroscience is aimed at uncovering the mechanisms behind behavior, which may include emotion, but is not limited to it. It can be seen as a branch of biology or psychology, but it is truly an interdisciplinary approach combining the fields of chemistry, computer science, engineering, linguistics, mathematics, medicine, genetics, philosophy, and physics. Technologies and methodologies of neuroscience can range from molecular and cellular approaches to brain imaging and behavioral analysis. Clearly, neuroscience is a broad field. Applied consumer neuroscience, a term defined by the current authors, can be described as a combination of neuroscientific, psychological, and traditional market research methodologies to better understand consumer behavior and nonconscious interactions with consumer products. More popular methodologies include a range of technologies from biometrics or autonomic nervous system measures (heart rate variability [HR], galvanic skin response [GSR], facial electromyography [fEMG]) to brain imaging (functional magnetic resonance imaging [fMRI]). However, the usefulness and validity of some technologies (fMRI or electroencephalography [EEG]) for consumer research is a hot topic for debate in the field and can often depend on cost and variability in the quality of the technology. The application on the perception of food products is even more problematic for practical reason (tasting presuppose movements that compromise the measurement).

The idea behind using applied consumer neuroscience (more popularly called "neuromarketing") to understand consumer preferences, attitudes, and behaviors is to delve into the nonconscious or noncognitive causes of these factors. The combined effect of these factors on product choice is mediated by emotional responses. Undoubtedly, it is still important to ask the consumer what they think. Conscious and nonconscious measures are providing very different answers (or should if done correctly). And so they should not be at odds with one another, but they should be providing added and synergistic information to help companies make better business decisions. If applied consumer neuroscience measures simply repeated results from explicit testing, then it would not be worth doing. The results must be synergistic and tell a more complete consumer story. In a larger viewpoint, it is possible to see how understanding consumer needs for products can help build better products (a top-down as opposed to a bottom-up approach to research). Understanding the consumers by using a combination of qualitative and quantitative research with applied consumer neuroscience, it could be possible to gain a deeper insight of the drivers of behavior and liking of consumer products.

5.2.2   Measuring Emotions Using Verbal Languages and Pictures

5.2.2.1   Explicit Measurements: Verbal and Visual Self-Reports

Verbal Self-Reports: Predetermined and Product-Specific Questionnaires
Among the explicit methods to measure emotions, verbal self-report is certainly the most common in psychology, and it is not weird that when emotions began to be measured in consumer science questionnaires were the first instrument used.

With some exceptions,[1] the classical approaches developed in psychology and psychiatry are not the most suited, for a number of reasons: first, because they are focused on negative emotions while in commercial product experience, positive emotions are predominant (Schifferstein and Desmet, 2010) and, in addition, because many of the terms included in psychological scales are not considered relevant by consumers to describe the emotions elicited by the tested product (Richins, 1997; King and Meiselman, 2010; Delplanque et al., 2012).

Several consumer studies have tried to develop a set of descriptors (a lexicon) that should represent the full range of emotions that consumers most frequently experience in consumption situations, suitable for more product categories (Richins, 1997; Laros and Steenkamp, 2005; King and Meiselman, 2010; Thomson and Crocker, 2013) or specific for a product category (Ng et al., 2013a; Spinelli et al., 2014a,b,c).

When measuring emotions in consumer research, it is a matter of discussion if we are measuring the actual "experienced" emotions, or the emotions associated with the product (Jaeger et al., 2013; Köster and Mojet, 2015; Spinelli et al., 2014a): while the former should be emotions elicited directly by the products, the latter are emotional associations that are elicited by the products through the mediation of individual and cultural memories, that record the emotions elicited in the personal product experience (individual memory) or through the media, eg, in the arts and in advertisements (cultural memory). That fact contributed to the great variety that we observe when emotions elicited by food products are measured using words (Jiang et al., 2014; Gmuer et al., 2015). This is true in psychology of emotions and is even wider in variability in consumer studies, where the lexicon was developed often with an inclusive approach, aimed at measuring both emotions and emotional associations. This information is very important if you are interested in understanding the product experience, at the price of a less orthodox and theoretically accurate perspective.

Using questionnaires, it is possible to profile products in a very similar way to the sensory profile, allowing the use of graphical representation of maps and even spider plots, depending on the scale used to measure emotional responses. In verbal self-reports, individuals are asked to express their emotions verbally, usually rating a battery of emotion items by using Likert scales, Intensity Scale, Best/Worst, Check-All-That-Apply or a direct scaling technique (eg, Bullseye; Thomson and Crocker, 2014).

When using a verbal emotional approach, it is important to first decide whether to use a standardized questionnaire or a product-specific questionnaire. Recently two standardized verbal questionnaires designed to collect consumer responses have been presented: the EsSense Profile™ and GEOS (Geneva Emotions and Odor Scale). In both of these questionnaires, the list of emotions has evolved from existing mood and emotion questionnaires and feedback from consumers to evaluate the appropriateness of the terms.

GEOS was originally developed to precisely study emotions associated with odors (Chrea et al., 2009; Porcherot et al., 2010). It was then applied with actual non-food products (Porcherot et al., 2013) and only recently on a drink product (Porcherot et al., 2015). GEOS consisted of 36 adjective emotional terms, grouped

---

[1] Kuesten et al. (2014), successfully applied Positive and Negative Affect Schedule (PANAS) developed by Watson et al. (1988) to measure consumer emotions associated with phytonutrient supplements with different aromas, a number of studies have proven that considering food and other commercial products.

into six classes by using factorial analysis. The new version of GEOS, ScentMove™ (Porcherot et al., 2010, 2012; Delplanque et al., 2012), consists only of six items, each labeled by a phrase identifying a class (pleasant feeling, unpleasant feeling, relaxation, refreshment, sensuality, and sensory pleasure) and illustrated by three words (nouns and adjectives). While it appeared very powerful in discriminating between odors and in the case of non-food product (hygiene, home, and personal care products), it did not discriminate between aperitif drinks in an ecological setting (Porcherot et al., 2015).

The EsSense Profile™, originally developed by King and Meiselman (2010), is the first predetermined questionnaire developed to measure emotions associated with food products in a consumer setting. The questionnaire is being used both on food names and food products and employs a list of 39 predetermined emotions and mood terms presented as adjectives, only 3/5 of which are negative. Recently a reduced version of 25 items has been validated (Nestrud et al., 2016). Respondents have to rate each emotion using a 5-point scale, answering a general question: "how does that food make you feel when you eat it?" (see King et al., 2013 for a detailed methodological discussion about the use and/or implement of this method).

EsSense was validated in a study that compared different food categories (name and actual products) and only more recently has been applied to discriminate within the same product category (Cardello et al., 2012; Manzocco et al., 2013; Ng et al., 2013a; Spinelli et al., 2014a; Gutjar et al., 2014). In this last decade, other approaches have appeared with the aim to develop product-specific questionnaires aside from the standardized questionnaires. These questionnaires were developed, thanks both to preliminary studies with consumers (interviews, focus groups, term selections, Repertory Grid Method) and literature, with the aim to elaborate a reduced list of emotions the most suitable for the product category (Thomson, 2007; Thomson et al., 2010; Ferrarini et al., 2010; Ng et al., 2013a; Spinelli et al., 2014a,b,c; Chaya et al., 2015).

Both of these approaches, predetermined and product-specific questionnaires, boast some advantages but also have some limitations. Predetermined questionnaires are cheaper and easier to use than product-specific ones, but they were found to be less discriminating for their general nature when compared to questionnaires developed for a specific product category (Ng et al., 2013a; Spinelli et al., 2014a,b,c). A standardized questionnaire such as the EsSense Profile™ has to include many terms to be certain it does not miss important emotional dimensions; this characteristic may imply some consequences, such as a fatigue/boredom effect on the respondents affecting the results (Ng et al., 2013a; Jaeger et al., 2013). Those are the main reasons why the authors suggest that a sample number of two products would be optimal (King et al., 2013), or eventually to modify with addition/ subtraction of the emotion terms to adapt the questionnaire to different class of foods or to the requirements of experimental design (King and Meiselman, 2010). Moreover, if some important emotions are missing from the list, respondents could give a higher rating to the emotions considered the closest ones but not really correspondent to the one felt.

On the other hand, ad hoc questionnaires reveal more detailed information about the product specificity and can furnish more fine-grained analyses, but they usually require additional preliminary work in developing the questionnaire.

Apart from the procedure chosen to develop questionnaires—predetermined or ad hoc—these approaches do not really differ in the final format; each presents to respondents a previously defined list of adjectives or nouns to select and/or rate to describe their emotional experience of the product. Adjectives are commonly recognized as the suited "labels" to indicate emotions, and they are usually preferred to nouns in the studies because they seem to be more easily associated with immediate emotional experience (Plutchik, 1980). However, the emotional lexicon includes also verbs, adverbs, nouns and interjections and may change considerably in different languages. What is not generally considered is that there is not a strict correspondence between emotions and the words used to indicate them. Often, in the absence of a specific word in a natural language, emotions are expressed using a sentence that paraphrases the meaning (Ogarkowa, 2013). In addition, the problem with many emotional words is that each has multiple and ambiguous meanings, depending on the contexts of the individual experience of each speaker (Kagan, 2007). In each text, in fact, the meaning of a word is selected by the context defined by the other words that surround it in a sentence and by the situation in which the sentence is included: the topic of the text fixes which semantic properties are "activated" and consequently have to be considered and which of them could potentially be activated, but they were not in that context (Eco, 1979; Eco, 1990). Thus, it should be considered that words need a context to be interpreted correctly, that is to say, in this case, to be interpreted in the way the researcher expects that they should be interpreted.

Questionnaires are particularly sensitive to this problem of ambiguity in wording (Belson, 1981). Jaeger et al. (2013) pointed out the problem of lack of understanding and misunderstanding in the EsSense emotion list and emphasized the absence of a meaningful context that could help to reduce ambiguity. In a recent study comparing EsSense Profile to a shortest version of 25-items, Nestrud et al. (2016) sustained that the meanings of some emotion words might vary within the context of a longer list. Presenting emotions organized in groups and not in a unique list has been tried as a way of addressing this problem (ScentMove™: Porcherot et al., 2010; Geneva Emotions Wheel: Scherer, 2005). Such a choice can be useful to help the respondent to better understand the task by clarifying the emotional area indicated. However, the use of complete sentences may provide greater clarity because of the inclusion of a context, in both a linguistic and situational sense. In fact, it can be reasonably supposed that the use of full sentences can be clearer than isolated words and able to reduce ambiguities. This is in line with some studies that noted that the emotional lexicon includes words that indicate an emotion only if included in a "feeling" context and not a "being" one: for example, the word "guilty" does not refer to an emotion in the sentence "I am guilty of something," but it indicates an emotional state in the sentence "I feel guilty" (Clore et al., 1987; Ortony et al., 1987).

Recently, Spinelli et al. (2014a) proposed EmoSemio, a protocol to develop a product-specific questionnaire able to solve some limitations of the current approaches and where the ambiguity of emotional words is reduced as much as possible.

The EmoSemio approach includes two steps:

1. Preliminary phase of questionnaire development with 20/25 one-on-one interviews conducted applying a modified version of the Repertory Grid Method. A semiotic methodology (Greimas and Courtès, 1979; Rastier, 1997; Violi, 1997) is used to analyze interviews and to identify emotions associated with products.
2. Collection of liking and consumer emotional responses using the EmoSemio questionnaire.

Reducing ambiguity is very important to obtain valuable and consistent results. Using a semiotic analysis of interviews allows exploration of the contextualized meanings of the words used to describe the experience by respondents; in addition, using full sentences instead of adjectives in the questionnaire help to contextualize the meaning.

The EmoSemio approach seems to be appropriate when the emotional profile of a specific product category is of interest, allowing a fine-grained analysis with relatively modest costs as to the benefits (20/25 interviews, that can be conducted also using an online video call device). In the last years, EmoSemio was applied in a commercial setting to a number of different products categories (cocoa and hazelnut spreads, coffee, cans of tomatoes, extra virgin olive oils, and laundry detergents) showing discrimination ability and suitability with different food and also non-food products (Spinelli et al., 2014a,b,c, 2015).

Pros and Cons of Using Questionnaires to Measure Emotions

From a methodological point of view, the choice of the verbal questionnaires has the advantage of measuring the subjective experience of a person during an emotion episode, with the rich nuances that only the language can express. Scherer (2005) noted that "while both nonverbal behavior (e.g. facial and vocal expression) and physiological indicators can be used to infer the emotional state of a person, there are no objective methods of measuring the subjective experience of a person during an emotion episode. Given the definition of feeling as a subjective cognitive representation, reflecting a unique experience of mental and bodily changes in the context of being confronted with a particular event, there is no access other than to ask the individual to report on the nature of the experience." The richness of language makes that approach very useful in sensory and consumer research, allowing profiling the products comparing blind and packaging conditions, discriminating products for specific emotions, and collecting information that can be used to communicate the emotional identity of the products. For an example of the results obtained with a verbal questionnaire (EmoSemio), see the case study presented in Section 5.3.3.

The limits of verbal measurements are linked first to the intrinsically ambiguous nature of language previously discussed. While language is the richest system we have to express meaning, it is as well ambiguous, and it cannot be taken for granted that people interpret words in the same way. Great attention and care should be paid to the development of the questionnaires verifying possible sources of ambiguity in language used. Second, it has been noted that explicit measurement (such as verbal) can guide the attention of the respondents, while some emotions may remain unconscious (Berridge and Winkielman, 2003; Köster and Mojet, 2015). However, if we consider with Scherer (2005) that subjective feeling is only one of the components of the emotion, what is measured with implicit or explicit methods could also be something different (two components of the emotion). Third, the use of language opens the problem of translation, such as in cross-cultural studies, or when a questionnaire developed in a language is translated into another one, and even when the study is conducted in two countries speaking the same language (Van Zyl and Meiselman, 2015). There is an ongoing debate on whether emotion words reflect universal psychobiological processes or, rather, culture and language-specific conceptualizations of emotional experiences (or both). The main and widest effort to understand the meaning of emotion terms was made by Fontaine et al. (2013) and all their coauthors who developed a multi-theory-based instrument to empirically assess the meaning of emotion words in different languages (GRID project). The research of this extensive study, conducted on 24 emotion words, in 27 countries representing 24 different languages, should be kept in mind in planning emotion cross-cultural studies, and also in the development of new questionnaires.

Visual Self-Report

Instead of relying on verbalizations or a list of emotion words, responses of visual self-reports are based on images representing different emotions or emotional states. The literature reported two main visual self-report instruments applied to food products: the Self-Assessment Manikin (SAM) and the more recent PrEmo. Recently, Collinsworth et al. (2014) proposed the Image Measurement of Emotion and Texture (IMET) as a new method applied on food and beverages products that utilizes a combination of images of emotions and texture (self-selected by the respondents in the case of emotion and provided by the researchers in the case of texture).

Self-Assessment Manikin Technique

SAM is a methodology for assessing emotional responses to stimuli based on non-verbal scales. Developed by Bradley and Lang in the 1980s, it utilizes pictorial scales (Fig. 5.1) based on the PAD (Pleasure-Arousal-Dominance) theory of emotion.

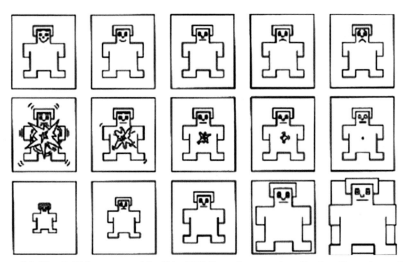

Figure 5.1 The Self-Assessment Manikin (SAM) used to rate the affective dimensions of valence (top panel), arousal (middle panel), and dominance (bottom panel). Image credit: from Bradley and Lang (1994).

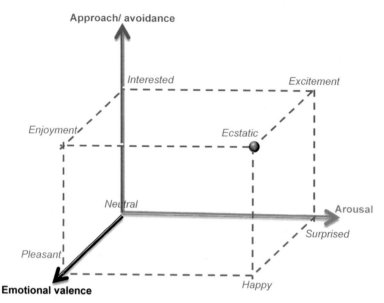

Figure 5.2 Three-dimensional mood mapping using PAD Theory of Emotion. Image credit: HCD Research.

The PAD model is a psychological model developed by Mehrabian and Russell to describe and measure emotional states using three numerical dimensions to represent all emotions (1974). These dimensions include pleasure, arousal, and dominance (Fig. 5.2).

Due to the numerical dimensions, the PAD model can be very useful for making statistical comparisons between situations and differentiating reactions. The PAD model assesses how pleasurable (hedonic appraisal), arousing (interest), and dominating (approachable) a stimulus or experience is. These assessments can be combined for more specific, three-dimensional emotional mapping. For example, if something were both pleasurable and arousing, it elicits happiness. If something were both unpleasant and un-arousing, it elicits boredom. While helpful to know that the experience was happy or boring, it is also important to consider the third dimension, dominance, as a differentiator. If a person was feeling both unpleasant and aroused, this may represent anger. However, a person may also feel unpleasant and aroused when experiencing fear. The difference can be seen in their assessment of dominance, or how willing they are to approach or withdraw from the stimulus. If they were more approaching, higher in dominance, they are experiencing anger, whereas if they were more withdrawing, lower in dominance, they are experiencing fear. Additionally, if we have two fearful experiences, we can compare statistically which is stronger, more intense, or less pleasurable based on the numerical scales. Using the three scales, it is possible to then create a three-dimensional emotional map of reactions to stimuli (Fig. 5.2).

The SAM methodology is a validated measure of psychological, emotional responses to stimuli that can include images, videos, colors, sounds, and words. As it relies on pictorial scales as opposed to the more traditional quantitative measure of

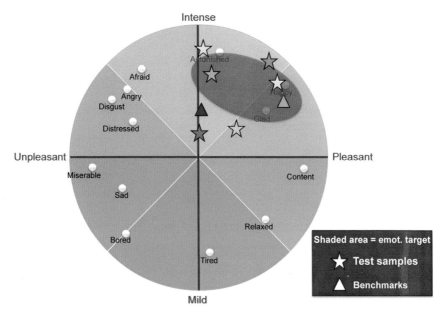

Figure 5.3 Two-dimensional mood mapping and consumer product testing (pleasantness versus arousal). The shaded region represents the emotional target for the product, in this case a positive and arousing, generally happy product concept. The stars represent test samples or prototypes. The triangles represent benchmarks for the test samples to perform against. Using this methodology, it is possible to map and compare test samples and benchmarks to aid in making go/no-go decisions on product changes and innovations. This approach can be easily used to compare flavors, colors, sounds, etc., associated with food or drink products and packaging.
Image credit: HCD Research.

numbered scales or descriptive scales (such as those ranging between strongly disagree–neutral–strongly agree), the SAM approach can be used across cultures and age ranges (including children), as well illiterate populations. This pictorial approach also claims to be able to assess noncognitive emotional responses, as the participant is not reading but reacting to imagery. However, while its "noncognitive" claim is somewhat debatable, the application of this methodology to consumer research is much more user-friendly than other emotional assessments. The SAM methodology assesses the emotional state of the person, specifically how they feel in response to a specific stimulus at the time of exposure while avoiding complications from language. This specificity makes this methodological approach very useful for consumer research. For example, it can be very useful in product development for assessing product attribute changes or for comparing with both internal and external benchmarks, such as current product or competitor products (Fig. 5.3).

Product Emotion Measuring Instrument

PrEmo is an instrument developed by Desmet (2003) to measure emotional responses using cartoon figures instead of words, applied with non-food and also food products

(Gutjar et al., 2014; Schifferstein et al., 2013). It includes 14 emotions, seven pleasant (ie, desire, pleasant surprise, inspiration, amusement, admiration, satisfaction, and fascination), and seven unpleasant (ie, indignation, contempt, disgust, unpleasant surprise, dissatisfaction, disappointment, and boredom). In the instrument, each of the 14 measured emotions is portrayed by an animation by means of dynamic facial, bodily, and vocal expressions.

It should be noted that the figures used are associated to the fictional world of tenderness of comics, and cartoons might condition the mood of the respondent while making a more pleasant atmosphere to perform the task (see also Köster and Mojet, 2015). The choice of another iconographic style for the drawings could potentially have a different effect on emotional experience. If this is true, it suggests that only relative differences within a set of products can be measured with this approach.

5.2.2.2   Implicit Measurements: Implicit Association and Emotive Projection Test

It has been noted that explicit method can lose some important information about emotions, which have a conscious but also an unconscious component. In addition, asking the respondent directly to think about his/her experience may induce a retrospective judgment of that experience, both based on rational or irrational thinking, which can modify the emotional response, at least the declared one. In addition, it can be difficult for people to state emotions that not correspond to the image that someone has of oneself, which brings them to modify their emotional responses (Köster and Mojet, 2015). For these reasons, it was suggested to adopt implicit measurement to investigate the unconscious emotional responses.

Implicit Association Test

The Implicit Association Test (IAT) is a methodology from social psychology developed by Greenwald et al. (1998) to detect the strength of a person's automatic association between mental representations of concepts. IAT has quickly become a valid and easy measure of implicit perceptions and psychological biases and has been used extensively in social psychology research of social stereotyping such as racism, sexism, ageism, etc. More recently the IAT has become popular for use in consumer science for measuring implicit preference for brands (Maison et al., 2004), affective responses to food (Mai et al., 2015), and cross-modal associations in product packages (Parise and Spence, 2012; Piqueras-Fiszman and Spence, 2011; Piqueras-Fiszman et al., 2012) and between pitch and basic tastes (Crisinel and Spence, 2009).

The IAT is a computer-programmed forced choice test, which measures the implicit association between concepts throughout the reaction time: the quicker the response, the stronger the association. This type of reaction time testing can be used to uncover the implicit or noncognitive association that consumers have toward brands and product attributes (such as package design, package communications, and sensory attributes such as taste, smell, color, and sound qualities). When applied effectively, IAT can be used to understand the strength of the emotional and perceptual effect of product communications on consumer product experience.

Emotive Projection Test

It has been noted that actually implicit methods and the physiological measures of reactions could be not completely implicit, because even if the subjects have no explicit power over their reaction and they are not explicitly told about the aim of the study, the experimental situation may lead the people to wonder what is the relationship between the measurement and the stimulation with food or food packages or food names and to make the connection (Köster and Mojet, 2015).

For that reason, a more indirect approach was recently suggested using a projection method. The test is based on the projections of the participants' positive and negative feelings on a set of photographs showed to them after the stimulation (eg, tasting the products or smelling an odor). This has the advantage of measuring mood effects of external stimulations without any explicit awareness of the relationship to its cause by the participating subjects (Köster and Mojet, 2015; De Wijk and Zijlstra, 2012). In a recent study, Mojet et al. (2015) found that a new emotive projective test (EPT) was useful to measure implicit emotion evoked by yogurts. In the EPT the subjects rated photographs of others on six positive and six negative personality traits after having eaten the yogurt. The results showed clear differences in the projected moods evoked by the yogurts in two of the three sets of stimuli; in addition, the differences as indicated by the results of the individual subjects were independent of their liking of and familiarity with the products. However, although liking itself was not correlated with the emotional effects in the emotive projection test, shifts in liking caused by consumption of the product (expected liking versus blind liking) did, indicating the emotional importance of pleasant surprise or disappointment in the confrontation between the expected and the actual experience of the product. These findings are in line with the discussion of the impact of expectations on emotions in Section 5.3.4.

5.2.3 Measuring Emotions Through Physiological Measures (Autonomic Nervous System)

The autonomic nervous system (ANS) is the part of the nervous system responsible for control of the bodily functions not consciously directed, such as respiration, heart rate, papillary response, and digestive processes. Measurement of ANS responses can help researchers understand unconscious reactions to stimuli to uncover nonconscious "thoughts" or reactions. Here we are using "unconscious" to describe the physiological actions of the ANS that are not consciously directed and "nonconscious" to describe the psychophysiological reactions to stimuli. These psychophysiological reactions have been found to be associated with emotional processes in academic literature (Kreibig, 2010).

In applied consumer neuroscience, the physiological ANS responses to stimuli are recorded externally via sensors. The most popular methodologies for ANS measurements in consumer research include facial electromyography (fEMG), heart rate variability (HRV), GSR, also sometimes termed "skin conductive response," and eye tracking (including pupillometry or measuring pupil dilation) (Fig. 5.4).

Facial Electromyography (fEMG) utilizes sensors placed on the skin to measure muscle activity of two main muscle groups associated with emotional reactions and

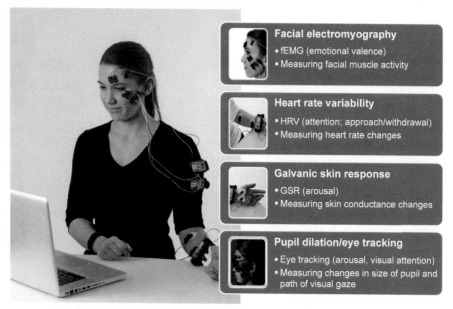

Figure 5.4 Autonomic measures: psychophysiological methodologies.
Image credit: HCD Research.

valence: the corrugator supercilii group, which is associated with frowning, and the zygomaticus major muscle group, which is associated with smiling (Magnée et al., 2007; Larsen et al., 2003). Emotional valence, a scale of mood state, ranges from more positive to more negative. Research has shown that muscle activity from the corrugator muscle, for example lowering the eyebrow, is involved in producing frowns, correlated with a more negative emotional valence. Activity of the zygomatic major muscle, involved in smiling, is positively correlated with positive emotional valence.

Heart Rate Variability (HRV) is the assessment of the variation in the time interval between heartbeats, measured by the variation in the beat-to-beat interval and can be recorded via electrocardiogram (ECG) sensors placed on the body (Williams et al., 2015; Quintana et al., 2014). HRV is inversely associated with motivation responses; it increases when focus and attention is low and decreases when focus and attention is high. For example, when problem solving, response selection and initiation will reflect HRV decrease as sympathetic outflow prepares the system for action; in contrast, vigilance—which is attention directed outward such as scanning or stimulus anticipation—is accompanied by bradycardia and HRV increase (Carroll and Anastasiades, 1978).

Galvanic Skin Response (GSR or SCR) is used to measure electrodermal activity (EDA), a property of the human body that causes continuous variation in the electrical characteristics across the skin. Skin resistance varies with the activity of sweat glands in the skin, controlled by the sympathetic nervous system, and skin conductance is an indication of psychological or physiological arousal (Gouizi et al., 2011).

Eye tracking measures the point of gaze (where one is looking) or the motion of an eye relative to the head and can also be used to measure pupil dilation or **pupillometry**, the measurement of pupil diameter. Eye gaze can be a valuable tool for assessing visual attention (different from cognitive attention). Pupil dilation responses reflect the resolving cognitive tasks; greater pupil dilation is associated with increased cognitive processing in the brain (Bradley and Lang, 2015). A recent study suggested that pupillometry can be used to measure emotional responses to food but not to qualify emotions; thus, a combination with a self-report measure is also requested (Lemercier et al., 2014). In addition, further studies are required to ensure that it is a measure of emotions and not just arousal.

Product packaging and marketing is designed for attracting attention and directing the consumer experience, even before the product is used. Recent studies using eye tracking have shown that front-of-package nutritional labels grab more attention than the traditional back of package nutrition facts (Graham et al., in press) and that gaze order can influence the overall package experience (Piqueras-Fiszman et al., 2013). By combining ANS measurement tools (fEMG, HRV, GSR, and eye tracking), researchers can assess the communications (graphics, messaging, etc.) for emotional impact and engagement (Kenning and Plassmann, 2008). Order of gaze and visual attention heat mapping can reveal how graphic and informational flow can affect packaging success, while other ANS measures (fEMG, HRV, GSR) can help one to understand the emotional and physiological response (Liao et al., 2012, 2015). On a study on consumer emotional responses to packaging of chocolate bars using both SAM and ANS measures, Liao et al. (2015) found that images generate an emotional response that can be measured by both self-report and physiological measures, whereas colors and typefaces generate emotional response that can only be measured by self-report measures. For that reason, the authors suggest that a joint application of self-report and physiological measures can lead to richer information and wider interpretation of consumer emotional responses to food packaging elements than using either measure alone.

5.2.4 Measuring Emotions From the Brain

While neuroscience is the study of the mechanisms behind behavior, cognitive neuroscience addresses the deeper questions of how psychological functions are produced by neural circuitry. Neuroimaging technologies, electrophysiology, and human genetic analysis combined with sophisticated experimental techniques from cognitive psychology have allowed neuroscientists and psychologists to address deeper questions of how human cognition and emotion are mapped, tracking, and mapping neural activity across the brain in search of patterns, or brain states, that correlate with psychological or cognitive processes.

The brain is an electrochemical organ, the electrochemical activity occurs in very specific ways that are characteristic of the human brain: brainwaves (Christoffersen and Schachtman, 2016). There are four categories of these brainwaves, ranging from the most to the least activity: beta, alpha, theta, delta. What do these different brain states mean? The methodological breakthroughs of neuroscience and technology, in particular the introduction of high-tech neuroimaging procedures, have enabled scientists to begin to describe and comprehend the neural processes which correspond to mental functions. Though, importantly, these measures do not "read minds." Neuroscientists use empirical approaches to discover neural correlates of subjective phenomena such as emotion. It is important to also remember that there is great redundancy and parallelism in neural networks. While activity in one group of neurons may correlate with a percept in one case, a different population might mediate a related percept if the former population is lost or inactivated.

5.2.4.1 Electroencephalography

EEG is a noninvasive method to record the electrical activity of the brain along the scalp for measuring brain states (Soloman, 2010; Nunez and Srinivasan, 2006). The recording is obtained by placing electrodes on the scalp with a conductive gel or paste, usually after preparing the scalp area by light abrasion to reduce impedance due to dead skin cells. In standard clinical practice, 19 recording electrodes are placed uniformly over the scalp (the International 10–20 System). In addition, one or two reference electrodes and a ground electrode are required. Additional electrodes (up to 256 electrodes) can be added to the standard set-up for increased spatial resolution for a particular area of the brain. EEG voltage signals represent the difference between the voltages at two electrodes.

Most of the available studies on emotion-specific EEG response have focused on EEG characteristics at the single-electrode level; however, emotional state is likely to involve circuits rather than any brain region in isolation. EEG-based studies of emotional specificity, with analyses at the single-electrode level, have demonstrated that asymmetric activity of the frontal regions (especially in the alpha) is associated with emotion. For example, Ekman and Davidson (1993) found that voluntary facial expressions of smiles of enjoyment produced higher left frontal activation. However, Coan and Allen (2003) found decreased left frontal activity during the voluntary facial expressions of fear. In addition to alpha band activity, theta waves at the frontal midline area have been found to relate to pleasant (as opposed to unpleasant) emotion (Sammler et al., 2007).

Emotional interpretations of EEG results are complicated and not yet fully understood. However, as previously mentioned, emotion is a complex process and may not be as simple as these studies suggest as some debate has arisen over the issue. Human beings are complicated creatures. One of the gravest mistakes that we can make in using neuroscience is the oversimplification of very complicated processes.

Despite the relatively poor spatial sensitivity of EEG (location), it possesses multiple advantages over some of these techniques. Hardware and personnel costs are significantly lower than those of most other techniques. EEG has very high temporal resolution (timing), on the order of milliseconds rather than seconds as seen in brain imaging technologies such as functional magnetic resonance imaging (fMRI) and near-infrared spectroscopy (NIRS). Though most movement must be kept a minimum, EEG is relatively tolerant of subject movement, unlike most other neuroimaging techniques where subjects must remain still for long periods of time.

EEG also has several limitations. Most important is its poor spatial resolution. EEG is most sensitive to a particular set of electrical potentials generated in superficial (outer) layers of the cortex, closest to the skull. Signals from deeper portions of the brain have far less contribution to the EEG signal. Unfortunately, most of the higher order psychological and emotional areas of the brain are located at deeper levels within the brain, such as the amygdala or hypothalamus. Additionally, EEG ERPs (event-related potentials, the electrical pulses representing brain activity) represent averages of thousands of neurons; a large population of cells in synchronous activity is necessary to cause a significant signal in the recordings. Therefore, EEG provides information with a large bias to select neuron types, and it generally should not be used to make claims about global brain activity. Unfortunately, the majority of consumer neuroscience emotional studies use global changes to suggest emotional shifts. Furthermore, the meninges (lying between the skull and brain), cerebrospinal fluid, and skull distort the EEG signal, obscuring its intracranial source and complicating location or neural structural–derived interpretation.

Inexpensive EEG devices exist for the low-cost research and consumer markets, making the technology more accessible and easier to use. Recently, a few companies have miniaturized medical-grade EEG technology to create versions accessible to the wider public. Some of these companies have even built commercial EEG devices retailing for less than USD$100. However, this has also led to less vigorous research designs using less rigorous tools that are not validated to the same degree as clinical EEG setups. In addition, these cheaper headsets lack both reliability and validity. These headsets utilize significantly fewer electrodes placed across the scalp (sometimes as low as two electrodes), resulting in poor resolution. Further, these headsets use dry electrodes, while clinical- or academic-grade EEG use gel-assisted electrodes to maximize signal. Dry electrodes can result in increased impedance and distortion of the measures through the skull and meninges.

Recently, Ma et al. (2014) used EEG to understand if tastes and claims on food packaging can have an effect on consumer perception of food product, ultimately affecting consumer decision-making. Sweet-tasting foods presented along with information on potential side-effects and diseases elicited more conflict than salty food, as reflected by a more negative ERP component. Additionally, chronic diseases (hypertension, osteoporosis, etc.) aroused a stronger emotional fear than acute diseases (diarrhea, skin rash, etc.). This study suggests that consumer evaluation of foods combine taste with risk information and health claims for making an ultimate purchase and appetitive decisions.

5.2.4.2 Neuroimaging

Neuroimaging includes the use of various techniques to directly or indirectly image the structure and/or function of the nervous system. The nervous system comprises the central nervous system (brain and spinal cord), while the peripheral nervous system comprises the nerves and ganglia outside of the brain and spinal cord. Due to cost and equipment constraints, most neuroimaging technologies are not feasible in consumer research. The neuroimaging tools most frequently used in applied consumer neuroscience include fMRI, and to a lesser extent newer tools like fNIRS. Other neuroimaging tools like MEG, PET, SPECT, etc. are less common in consumer research, predominantly used in clinical and academic settings where such expensive equipment and medical staff are available.

Functional magnetic resonance imaging or functional MRI (fMRI) is a neuro-imaging procedure using measuring brain activity by detecting changes in blood flow (Huettel et al., 2009). Interpreting fMRI results rely on the idea that cerebral blood flow is correlated with neuronal activation. Such that if an area of the brain is in use, blood flow to that region also increases in that area. Analysis of fMRI is performed via measuring the changes via blood oxygen level–dependent (BOLD) contrast using the hemodynamic response to map neural activity, measuring for changes in magnetization between oxygen-rich and oxygen-poor blood. As it does not require invasive techniques like injections or surgery, it is relatively feasible for consumer ruse. However, this method can be easily corrupted by noise (from the noisy machine and environment, random brain activity, etc.) and statistical biases. To make up for the variability, fMRI studies must repeat a stimulus presentation multiple times. fMRI brain activity is recorded as voxels, a three-dimensional space imposed on an averaged brain image to represent the activity of millions of neurons and billions of synapses depending on the size of the voxel and mapping to identified Brodmann areas for functional interpretation.

While fMRI is very good for identifying neural locations and structures (spatial resolution), it is not as useful for temporal resolution. Recorded fMRI signals can lag behind neuronal events by several seconds. The temporal resolution is dependent on the type of brain processing being recorded as there can be a broad range of time required for different types of processing. Additionally, as fMRI testing procedures can take a significant amount of time, subjects may move their heads over time, resulting in drift of signals from baseline over time, affecting the accuracy of spatial recordings. Another potential issue in analyzing and interpreting BOLD signals is that the BOLD response does not separate for feedback and feed-forward networks; both inhibitory and excitatory input to and from neurons are summed into the BOLD signal, which may cancel each other out. Additionally, though the strength of the BOLD signal is supposed to suggest an amount of brain activity, it does not necessarily reflect behavioral performance, particular for repeated tasks where initial measures will be higher. but as the subjects get practice, amplitudes may decrease while performance remains high. While BOLD responses can be compared between subjects for the same brain regions, they cannot be compared across regions or tasks since density of neurons and therefore activity may be different per the region.

This means that much care must be taken in designing a well-controlled fMRI experiment, minimizing noise. When used correctly, it is possible to detect correlations between brain activations and tasks in subjects and to correlate a subject's emotional experience with a high level of accuracy. However, even with the best and most controlled experimental design, it is not possible to control and constrain all background stimuli and noise-producing neural activity independent of the experimental manipulation.

Some experiments have shown the neural correlates of peoples' brand preferences. McClure et al. (2004) used fMRI to show the dorsolateral prefrontal cortex, hippocampus, and midbrain were more active when people drank unblinded Coca Cola product as opposed to when they drank blinded Coke product. Other studies have shown the brain activity that characterizes men's preference for sports cars (Erk et al., 2002).

However, these studies suffer from a reverse inference problem, the logical fallacy of affirming what you just found. With regard to the brain and brain function, it is seldom that a particular brain region is activated solely by one cognitive process (Plassmann et al., 2012). Additionally, many scholars have criticized fMRI studies for problematic statistical analyses, often based on low-power, small-sample studies. In one real but satirical fMRI study, a dead salmon was shown pictures of humans in different emotional states while being scanned in an fMRI (Bennet et al., 2009). The authors, graduate students who presented their work at the Society for Neuroscience annual meeting, provided evidence, according to two different commonly used statistical tests, of areas in the salmon's brain suggesting meaningful activity. The study was used to highlight real analysis concerns over fMRI studies and the need for more careful statistical analyses in fMRI research.

With these constraints and concerns over fMRI, **functional near-infrared spectroscopy (fNIRS)** measurement seems to have strong potential for applicability in consumer research. Similar to fMRI, fNIRS is a noninvasive optical brain imaging technique that investigates cerebral blood flow as well as the hemodynamic response in a local brain area during neural activity (Jackson and Kennedy, 2013). fNIRS involves the irradiation of near-infrared light by diodes near the participants' head and scalp. This light is then detected via optodes and measuring its scattering position (Villringer et al., 1993). Oxy and deoxyhemoglobin absorb this light, so changes in their flow can be assessed as a correlated quantification of neural activity through calculations for hemodynamic response. Accuracy, however, is dependent on the number and distance between the diodes and optodes and further processing methods for artifact correction and algorithmic calculation.

fNIRS is a newer technology and not as widely used or accepted as fMRI, but it can provide similar information regarding consumer or participant emotional state. Recent fNIRS studies have shown lateralized hemodynamic responses in the prefrontal cortex in response to emotional cues correlated to explicit emotional appraisals, suggesting it is possible to use fNIRS technology to assess emotional states (Balconi et al., 2015). Studies on consumer choice such as that by Luu and Chau (2009) examined product preference using fNIRS, recording the amount of activity in the prefrontal cortex (associated with planning and decision-making). Participants were asked to look at two different drinks and to mentally evaluate their preference. Their results showed that subjective preference could be measured with 80 percent accuracy for activity in the prefrontal cortex to stated consumer choice. These studies, Balconi et al. and Luu and Chau, suggest that fNIRS may be a useful tool for measuring consumer emotions and preference.

5.2.5   Measuring Emotions Through Behavior

Observational research has had a long history in the study of psychology. It is a type of correlational (ie, nonexperimental) research in which a researcher observes ongoing behavior, more prevalent in social sciences and marketing. Observational or field research involves the direct observation of phenomena in their natural setting, meaning no intervention by the researcher. It is simply studying behaviors that occur, unlike the artificial environment of a controlled laboratory setting. Importantly, in naturalistic observation, there is no attempt to manipulate variables. It permits measuring what behavior is really like.

Quantifying behavioral actions involve categorizing and coding the behaviors. It is a measure of the behavior directly, not reports of behavior or intentions. The main disadvantage is it is limited to behavioral variables. It should not be used to study cognitive or affective variables. Another disadvantage is that observational data may not be generalizable.

In marketing research, the observational techniques more frequently used are vocal, facial, and whole-body behavioral/behavioral coding. Here we will focus on the second and the third, which have been used in food-evoked emotions studies.

5.2.5.1 Facial Coding

Based on the emotional theories of Carl-Herman Hjoortsjjo (1969) and later Paul Ekman (Ekman and Friesen, 1975; Ekman, 1989), facial coding is a system to taxonomize human facial movements into emotional categories. Originally coded by human experts, a facial active coding system (FACS) was established as a computed automated system that detects faces in videos to minimize subjectivity and variability. The automated system extracts geometrical features of the faces from captured video recordings and then produces temporal profiles of each facial movement including categorization of the calculated facial emotions and intensities (Fig. 5.5). Coding involves action units (the facial features that move), action descriptors (the direction of the movement), and intensity scores (the extent of the movement).

Facial coding has become a popular method for collecting the emotional reactions in "neuromarketing" work. Being able to capture consumer responses via quick videos

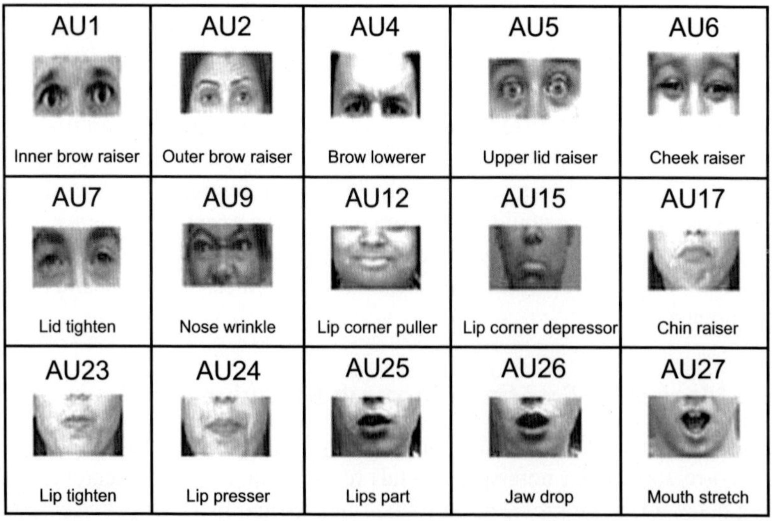

AU1	AU2	AU4	AU5	AU6
Inner brow raiser	Outer brow raiser	Brow lowerer	Upper lid raiser	Cheek raiser
AU7	AU9	AU12	AU15	AU17
Lid tighten	Nose wrinkle	Lip corner puller	Lip corner depressor	Chin raiser
AU23	AU24	AU25	AU26	AU27
Lip tighten	Lip presser	Lips part	Jaw drop	Mouth stretch

Figure 5.5 Examples of some action units extracted from Cohn and Kanade's database (Cohn et al., 1999).

(like surveillance in stores) and webcams, it is certainly an attractive idea to marketers. It seems cheap and fast. But as with all of the methodologies discussed, it has advantages and disadvantages.

First, the very theories that facial coding is based on are still hotly debated. Within this chapter, we have already discussed a few different theories of emotion, using different methodologies to assess emotions. Facial coding begins with Charles Darwin's theory that emotions are biologically determined and universal to human culture as he described in *The Expression of the Emotions in Man and Animals* (1872). Adding to this theory, Ekman found a high agreement across members of diverse Western and Eastern literate cultures on selecting emotional labels that fit facial expressions exhibited by actors. Ekman's famous test of emotion recognition, the Pictures of Facial Affect stimulus set, was published in 1976, consisting of 110 black and white images of Caucasian actors portraying the six universal emotions plus neutral expressions. These universal facial expressions included anger, disgust, fear, happiness, sadness, and surprise. A possible seventh universal emotion is contempt, though this has been debated as the evidence is less clear.

Some of the criticisms of Ekman's work are based on experimental and naturalistic studies by several other emotion psychologists that have not found evidence to support Ekman's categorization or universality of facial expressions (Chiao, 2015; Jack et al., 2009, 2012; Yan et al., 2015; Gendron et al., 2014). Additionally, practitioners have encountered some difficulties in its application given the categorical nature and variance in available technology (webcams, single low-definition video). It has also been suggested that the sensitivity of facial coding may not be as good as other methodologies (electrophysiological, etc.).

Facial expressions have also been used to measure emotional responses related to food. For example, positive facial expressions of newborns toward liked (sweet) and the negative expressions toward disliked (bitter) basic tastes have been established by Steiner (1973). Automated tools using FACS have been used for more diverse, universal facial expressions. Using these computerized tools to measure consumer responses to orange juices, researchers have found that happy expressions are surprisingly not related to liking scores, in contrast to neutral, angry, and disgusted expressions (Danner et al., 2014a,b). Researchers found that liking was positively associated with neutral facial expressions and negatively associated with facial expressions of sadness, anger, and scared in a study on breakfast drink (DeWijk et al., 2014) and food odors (He et al., 2014). These results support previous findings of a pilot study on children conducted by Zeinstra et al. (2009), who suggested that facial expressions can be suitable to measure dislike but not to measure various gradients of food acceptance. In addition, the facial expression of surprise was found to be both positively correlated with liking (He et al., 2014) and negatively (De Wijk et al., 2014).

Such contrasting results to what one would expect (happy facial reactions not indicating liking), complicate the use of facial coding methodology for assessing consumer liking of products. It was noted that facial expressions of happiness are rarely displayed when one is alone, suggesting that happy emotional reactions may have more of a social purpose than appetitive purpose (Parkinson, 2005). In addition, the fact that the basic emotion theory includes only one positive emotion (happiness), considering that surprise can be positive or negative depending on the situation, is another critical point in the application of this approach to measure consumer emotions elicited by food. All these findings suggest that facial coding is not suitable when the interest is to discriminate between products that are supposed to have a positive emotional performance (highly liked products).

Further, stronger facial expressions to disliked foods compared to liked foods are very quickly detected at the first visual encounter (De Wijk et al., 2012). The timing of facial expressions to food presentations (in sight and taste) may be very important to understanding the emotional impact and appraisal of food. Early reactions to odor presentations, like raising the eyebrows and opening the eyes wide, have been found to be related to the detection of a novel or unexpected stimulus, associated with increased alertness and attention, followed by an appraisal of the food for pleasantness (Delplanque et al., 2009; He et al., 2014).

The use of facial responses is technically more challenging than questionnaires and applications are therefore more suitable for laboratory than for Central Location Test. The situation in measuring food-related emotions is also complicated by the mouth movements that hamper the registration of emotion during tasting by the automatic registration systems (Mojet et al., 2015). Due to these limitations, facial behavior seems more suited to measure responses to food names or images than to tasted food (Köster and Mojet, 2015).

5.2.5.2 Behavior Coding

Behavior coding is the systematic assignment of codes or labels to the overt, observed behavior to analyze such behavior. It is typically evaluated via video and audio recordings in observational studies, but it can also be done live by expert coders. The behaviors observed through the video or audio are coded and then further analyzed for patterns, frequencies, etc. (categorical codes, durations computed from onset and offset times, straight transcripts of speech, informal comments, and so on).

Behavioral codes lie along a continuum. Implicit (sometimes called "subjective") codes are at one end of the continuum and explicit (sometimes called "objective") codes are at the other end of the continuum. Implicit codes do not require the observer to see particular behaviors; explicit codes do require this. Implicit criteria can be subjective, allowing coders to determine the code based on their own judgments of what behavior is being expressed; coders can take individual differences between participants into account. Explicit criteria force coders to determine the code based on whether a particular behavior was expressed; individual differences between participants and particulars of the situation must be ignored. With an implicit code for "pushing," for example, coders use their own judgment to decide whether they successfully and purposefully pushed a button on a product. With an explicit code for "pushing," coders must use explicit criteria such as whether the consumer's finger touched the button, whether the button was fully pressed, or whatever the predetermined definition of "pushing" was described as. With an implicit code for "negative affect," coders use their own judgment to decide whether a consumer feels distress or anger. With an explicit code, coders must use explicit criteria such as whether the consumer's brows were knit, lip was jutted, mouth was in a square shape, or crying/tears were expressed. Implicit and explicit codes can be equally reliable (in terms of inter-rater reliability and consistency of participants' responses) and equally valid (meaning that the codes reflect the behaviors we intend to measure). The benefit of an explicit code is that it is possible to know exactly what coders scored. Nonbehavioral codes can also be scored: for example, import information about participant demographics, the observational setting, various conditions, and independent variables.

Behavioral coding cannot by itself inform about consumer emotional response, but when paired with other measures (voice analysis, facial coding, etc.), it can be a powerful tool for interpreting the holistic consumer interaction with a product. Problems can arise if video recording arrangement does not record all relevant behaviors or if behavioral coding schemes are not informative. Therefore, research design and planning are paramount to successful behavioral analysis.

5.3 Emotions in the Product Experience: From the Product to the Packaging (and Back)

Although experienced as a whole, a product is a combination of very different items: the physical object characterized by a specific sensory identity, the package, the brand name and the marketing mix, and the context of usage or consumption (Schifferstein, 2010, 2015). Each of these items may elicit emotional response and may have different meanings to consumers. Every different aspect that constitutes the product, including the context and the interaction with the person that is experiencing it may create expectations that can be confirmed or disregarded provoking emotional responses.

Desmet and Schifferstein (2008) identified five main sources of food emotions: emotions elicited by (1) sensory properties, (2) experienced consequences, (3) associated consequences, (4) personal or social meanings, and by (5) behavior of agents involved. What it is apparent is that at different stages, different characteristics of the product prevail, and the emotions evoked can vary (Schifferstein et al., 2013; Labbe et al., 2015). When we consider a product on a shelf in a supermarket, for example, it is clear that the packaging plays a very important role, communicating the brand identity and creating expectations for both its sensory and branding aspects. These expectations could be confirmed or not by the actual experience of the product, eliciting specific emotions, such as pleasant surprise (when the actual experience is better than the expected one) or disappointment (when the actual experience is worse than the expected one). On the other hand, products can elicit specific emotions for their sensory characteristics not supported by the brand/packaging, causing problems in the product performance: for example, a product that elicits relaxation and calm paired with a communication that elicits energy and tension. In addition, considering repeated exposures, a product can elicit indifference or boredom, which are recognized as frequent causes of product failing (Köster and Mojet, 2007).

Aligning the emotions elicited by the product and the pack with branding in a consistent and congruent manner augments and strengthens the brand greatly (Lindstrom, 2005;

Krishna, 2012). This so-called "SensoEmotional optimization" (Thomson, 2007) can be developed in two ways, depending on the chosen starting point:

- Brand-first strategy: aligning the sensory characteristics of the product with the defining emotional characteristics of the brand;
- Product-first strategy: determining in the first place whether or not any of the defining sensory characteristics of the products are emotionally active, and on this basis, completely (re) building or otherwise reshaping or augmenting the brand.

5.3.1   Sensory Drivers of Emotions, Liking, and Choice

The rationale behind that approach is the fact that different sensory characteristics can drive different emotions in a specific product or product category (eg, a sweeter taste can elicit happiness).

Thomson et al. (2010) suggested that the conceptual profile of an unbranded product arises via three sources of influence: (1) category effect (how consumers conceptualize the product category), (2) sensory effect (how the sensory characteristics of a particular product differentiate it from other products in the same category), and (3) liking effect (the disposition of consumers to the category and how much they like a particular product).

The familiarity effect should be added (4) to those, considering the weight that this aspect has proved to have in consumer experience of products. For example, some products have a strong sensory signature that is recognizable by the consumer even when the product is presented in blind conditions producing a (learned) association of the sensory signature with emotions or meanings associated with branded product (Thomson, 2007). In addition, the more familiar product (that is to say, with a more familiar sensory profile) is usually the product that we expect to elicit more positive emotions, especially related to the area of reassurance.

To present, few studies have investigated the relationship between sensory and emotions in depth. In an inspiring and seminal study, Thomson et al. (2010) presented nine sensorially differentiated dark chocolates, unbranded, to consumers. The study showed that specific sensory characteristics were associated with emotional conceptualizations: for example, cocoa was associated with powerful and energetic, bitter with confident, adventurous and masculine, creamy and sweet with fun, comforting and easygoing. Unfortunately the liking measurement was not reported, so it is not possible to know if the identified sensory driver of emotions were different from the sensory drivers of liking. When liking was measured in other studies (Bhumiratana, 2010; Ng et al., 2012, 2015; Jager et al., 2014; Spinelli et al., 2014a; Chaya et al., 2015), it was apparent that these three aspects—sensory, emotions and liking—are interrelated, but also that emotional measurement captures information not captured by liking measurement. Many studies emphasize that emotions may discriminate beyond liking, as was found using self-report studies (Spinelli et al., 2014a, 2015; Ng et al., 2013a,b; King et al. 2010) and recently using implicit measurement (Mojet et al., 2015). This demonstrates very clearly that collecting emotional responses can help in interpreting product performance, especially when no significant differences in liking between products are found. This fact could be explained as an effect of "redundancy" on the dimension of valence provoked by emotions: measuring not only one but many emotions, both positive and negative, it is possible to interpret the relative differences between the emotional performance of the product. This is not surprising if we consider that valence (positive or negative) is the main dimension underlying the meaning of an emotion (see in Section 5.3.4). The segmentation of consumers on the basis of their preference could give deeper insight into that issue, allowing the more detailed investigation of the close relationship between preference for specific sensory properties and emotions (Bhumiratana, 2014; Köster and Mojet, 2015).

Further investigation is needed to verify more in depth the relationship between sensory properties, emotions, and liking. This information could be of great interest in laying the foundations of the so-called "sensory marketing" (Krishna, 2010, 2012).

Recent findings have shown that food-evoked emotion scores are better predictors of food choice than liking scores alone. Gutjar et al. (2014) and Dalenberg et al. (2014) found that evoked emotions predict food choice better than perceived liking alone in the case of breakfast drinks presented blind, and that the strongest predictive strength was achieved by the combination of evoked emotions and liking in both cases. In a further study, Gutjar et al. (2015) found that considering separately the valence and the arousal dimensions of the emotions, liking and valence together had the strongest predictive value for product choice based on products' taste, while the combination of liking, valence and arousal had the strongest predictive value for package-based choice. Further research is needed to verify the predictive value of the model in the case of other product categories.

5.3.2 Sensory and Branding: the Impact of Expectations on Emotions

Brand perception, in terms of associated emotions and meanings, has always been routinely measured in marketing studies (Bagozzi et al., 1999; Richins, 1997). However, the emotions elicited by sensory and branding in combination have rarely been studied. These two faces of the product—branding/packaging and sensory—have traditionally remained separate fields of, respectively, marketing/communication and product development. Emotions are the link between these two sides, characterizing the product experience in its entirety.

Investigating the emotions elicited by a food product, considering separately its intrinsic sensory characteristics (blind condition), its packaging (expected condition), or both its intrinsic sensory characteristics and its packaging sensory characteristics and branding (informed condition), can give a deeper insight into product perception helping companies in the design and optimization of products that meet consumers' expectations.

A mismatch between incoming sensory information and expectations can lead to changes in product acceptability (Cardello, 2007). The role of manipulation of information in creating expectations that guide liking and sensory perception has been extensively studied (Caporale et al., 2006; Siret and Issanchou, 2000; Caporale and Monteleone, 2004; Cardello, 2003; Deliza and MacFie, 1996; Kähkönen, Hakanpaa and Tuorila, 1999; Kähkönen and Tuorila, 1996, 1998; Kahkonen et al., 1996; see also Cardello, 2007; Tuorila et al., 1998; Tuorila et al., 1994 for an overview). Recently, a growing number of studies have focused specifically on the impact of packaging in creating expectations (see Piqueras-Fiszman and Spence, 2015 for a review). Packages, as a result of their own sensory properties (colors, shapes, etc.), can significantly contribute to build expectations toward the product, influencing liking and sensory perception (Schifferstein et al., 1999; Ares and Deliza, 2010; Becker et al., 2011; Carrillo et al., 2012; Delgado et al., 2013; Labbe et al., 2013; Lange et al., 2000; Mizutani et al., 2012; Piqueras-Fiszman and Spence, 2012; Piqueras-Fiszman et al., 2013; Schifferstein et al., 2013; Schifferstein and Spence, 2008) and also emotions (Ng et al., 2013b). Furthermore, packages can shape expectations since they are an expression of the branding. Brands are perceived by consumers as everything associated with them: in fact, a brand is defined as a "bundle of information" (Riezebos, 1994) representing a cluster of knowledge, experiences, and emotions that are stored in memory (Van Dam and Van Trijp, 2007). Branding is not a factor that has received much attention in the sensory and consumer science field, with some exceptions (Jaeger, 2006; Li et al., 2015). Several studies have measured the combined effect of brand with other extrinsic cues (eg, price, package, name) on product evaluation, without aiming to disentangle their relative impact (Di Monaco et al., 2004; Guinard et al., 2001; Lange et al., 2002; Mueller et al., 2010; Varela et al., 2010). Only a few attempts have been made to investigate the relative influence of branding and packaging on liking (Mueller and Szolnoki, 2010). In fact, packages have often been specially designed for a study and manipulated to create experimental design based on the control of some variables. When existing commercial packages have been used, or the manipulation concerned more variables (eg, the picture/ image on the label), descriptive analysis (Murray and Delahunty, 2000a,b) and semiotic analysis (Ares et al., 2011; Piqueras-Fiszman et al., 2011; Spinelli et al., 2015) were applied in pre-studies to map the differences among packages, with the aim to identify which characteristics would contribute to which expectation.

On the other hand, few studies have been conducted to investigate emotions elicited by the product (via its sensory properties) and its packaging/brand separately. Thomson and Crocker (2015) studied the brand-product consonance in dark chocolate and in the Single Malt Scotch Whisky category, comparing the conceptual unbranded profile to the brand profile. In addition, the unbranded conceptual profiles of whiskies were also compared to the conceptual profile of five Scottish tartans with the aim of examining the effect of ad hoc variations in color and formality of design on consonance with the conceptual profile of the whiskies. Ng et al. (2013b) explored the relative roles of sensory and packaging cues on consumer conceptualizations (emotional, abstract, and functional) in the case of commercial blackcurrant squashes. Three different experimental conditions were considered: (1) blind (consumers tasting the product blind), (2) pack (consumers viewing the packaging), and finally (3) informed (consumers tasted and viewed the packaging concurrently). The results indicated that intrinsic sensory properties seemed to have a stronger association with consumer liking and emotions, whereas extrinsic product characteristics seemed to have a stronger association with abstract/functional conceptualization. Product configurations of the informed and blind conditions for liking and emotional conceptual profiles are closely aligned, suggesting that liking and emotional responses were influenced more by the sensory properties (blind condition) than the packaging cues of the products. On the other hand, product configurations of the informed and pack conditions for abstract/

functional conceptual responses were closely aligned, suggesting abstract/functional conceptual responses were more influenced by packaging cues (eg, "old-fashioned").

5.3.3   Case Study: Comparing Emotional Performance in Unbranded and Branded Food Products Measuring Expectations

In a recent study conducted by the Sensory Unit of University of Florence (Spinelli et al., 2015), emotions were studied in relationship with expectations among 120 Italian consumers who regularly ate chocolate and hazelnut spreads. They were asked to rate liking and to identify their emotional responses to six hazelnut and cocoa spreads using a questionnaire developed specifically for the product category, the EmoSemio questionnaire (Spinelli et al., 2014a), under three conditions: blind, expected, and informed (Fig. 5.6). Six commercial hazelnut and cocoa spreads were selected to represent the range of sensory and brand variability in the Italian market segment, using at these aims descriptive analysis (see Spinelli et al., 2014a) and semiotic analysis of product packages (Floch, 1990, 1995; Arès et al., 2011; Smith et al., 2010).

The study revealed differences among emotional profiles within the same product category and showed that they arise primarily from their sensory differences, confirming previous findings (Gibson, 2006; Chrea et al., 2009; King and Meiselman, 2010; Porcherot et al., 2010, 2012bib_Porcherot_et_al_2012; Thomson et al., 2010; Ng et al., 2013a,b).

In addition, this study revealed the impact of expectations on emotions, showing a correlation between liking (blind, expected, and informed) and emotions.

Significant differences were found in consumers' overall liking for the products under blind, pack, and informed conditions ($p \leq 0.0001$). Furthermore, expected liking for G, T, and E was significantly higher than when they were assessed blind (negative disconfirmation). In contrast, liking for product B was higher when presented blind than in the original package (positive disconfirmation), (Table 5.2).

In agreement with previous studies, these findings showed that product packages can generate higher or lower expected liking scores than the product presented blind. However, for the products in this study, liking was mainly influenced by perceived sensory characteristics of the product rather than by brand perception, which could only reinforce an already positive performance: in fact, a significant difference between

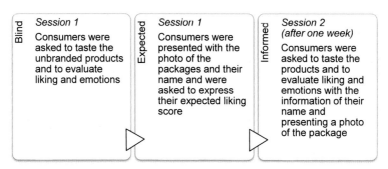

Figure 5.6 Experimental design of the study.

Table 5.2 Differences Between Blind (*B*), Expected (*E*), Informed (*I*) Liking, and Corresponding Probabilities (*p*) Tested Through Paired *t*-Test

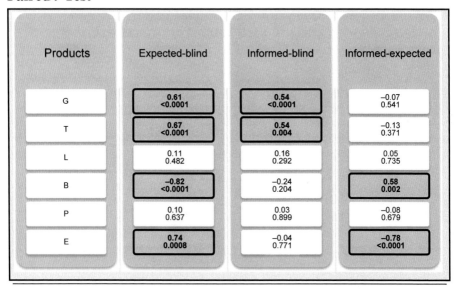

Products	Expected-blind	Informed-blind	Informed-expected
G	**0.61** **<0.0001**	**0.54** **<0.0001**	−0.07 0.541
T	**0.67** **<0.0001**	**0.54** **0.004**	−0.13 0.371
L	0.11 0.482	0.16 0.292	0.05 0.735
B	**−0.82** **<0.0001**	−0.24 0.204	**0.58** **0.002**
P	0.10 0.637	0.03 0.899	−0.08 0.679
E	**0.74** **0.0008**	−0.04 0.771	**−0.78** **<0.0001**

A significant difference is indicated in **bold**. (Spinelli et al., 2015).

the informed and the blind liking (I−B) was found only in the case of the most liked products in the blind condition (product G and T), while in the other cases, liking did not change in the two conditions, despite expectations. An assimilation effect following a negative disconfirmation was detected for products G and T only. For these products, the informed liking was significantly higher than the blind one. Informed and expected liking scores (I−E) of G and T were compared and significant differences were not found for either product, suggesting that consumers completely assimilated toward their expectations. The positive disconfirmation for product B did not result in an assimilation (no significant difference was found between informed and blind liking scores). Furthermore, informed liking for product B was higher than the expected one. This fact, in addition to the positive disconfirmation (E − B) suggested a problem in the package/brand. In contrast, expected liking for product E was higher than both informed and blind liking; although the brand was well known, the product was not liked.

Interestingly, in passing from a blind to an informed condition, positive emotions increased only for the products with a complete assimilation toward expectations (G and T), while some negative emotions decreased (Table 5.3). Moreover, some positive emotions increased only in one product but not in the other one: *curiosity*, *surprise*, and *generosity* increased only in the case of product T, while *childhood happy memory/reminiscence*, *amusement*, *energy*, and *relax* increased only for product G. This result supports the idea that the two products had different emotional profiles, built on a sensory identity and supported and amplified by branding/packaging.

Table 5.3 Comparison Between Emotions in Blind and Informed Conditions.

Valence	Emotions	Products					
		G	T	L	B	P	E
Liking		Complete assimilation			PD		ND
Pleasantness	Surprised		↑				
	Curious		↑				
	Energetic	↑					↓
	Sensual	↑	↑				
	Amused	↑			↓		
	Merry	↑	↑				
	Cuddled	↑	↑				
	Tender	↑	↑				
	Happy	↑	↑				↓
	Happy memory	↑		↓	↓		
	Generous		↑				
	Gratified	↑	↑				
	Satisfied	↑	↑		↓		
	Relaxed	↑					
	Anti-stress	↑	↑				
	Secure	↑	↑				
Unpleasantness	Guilty	↑			↓	↓	
	Indifferent	↓	↓	↓			
	Bored	↓	↓				
	Disappointed						
	Sad						
	Neglected					↓	
	Annoyed		↓				

Downward (↑) and upward (↓) arrows indicate the significant differences between informed and blind evaluations: ↑the emotions that were higher and ↓ the emotions that were lower in the informed condition compared to the blind condition. (Spinelli et al., 2015). *PD* and *ND* indicate, respectively, positive and negative disconfirmation in liking, not followed by assimilation.

Interestingly, our findings showed that the emotion of surprise did not discriminate among products in the blind condition, while it significantly discriminated in the informed condition. Moreover, changing the condition from the blind to the informed, the emotion of surprise, together with curiosity, became a relevant emotion in discriminating two positive emotional performances (G and T). In fact, comparing the blind and the informed results for the two products, the differences between G and T increased considerably and moved from the dimension of valence in a blind condition, to the dimension of novelty in an informed condition: from positive versus negative emotions (*happy memory*, *happy*, *gratified* versus *indifference*) to *curiosity* and *surprise* versus *happy memory of the childhood*. The emotion related to *happy memories* significantly increased for product G, and the gap between the two products for this emotion increased substantially. G was the oldest product in the market, the one that is recognized as part of the personal experience by the consumers and furthermore has an old-fashioned-style packaging; communication of product G was focused on memory, thanks to a mix of tradition and modernity with an effect of vintage. On the other hand, product T elicited more *curiosity* and *surprise* in the informed condition than in blind, possibly reflecting a positive unexpected experience that stimulated interest, suggesting that consumers were not familiar with this specific product even if the brand was well known.

Products B and E had a more negative emotional performance in the informed condition than when presented blind: three positive emotions (related to *amusement, happy memory*, and *satisfaction*) and one positive emotion (*happiness*) were, respectively, significantly lower when the products were evaluated associated with its packaging/branding. This suggests that a disconfirmation (both positive or negative) in liking not followed by assimilation is associated to a worsening of the emotional performance of the product. It is worth noting that for these two products, the informed liking did not significantly differ from the blind one, but some positive emotions were significantly less intense in the informed condition than in the blind condition.

When products did not differ in liking in the three conditions (L and P), very few significant differences in the emotional profile were found. However, emotions gave additional information to liking, suggesting in these cases a slight positive impact of branding/packaging, with the decreasing of some negative emotions (*indifference* for product L and of the *feeling neglected* and *guilty* for product P).

A higher number of negative emotions decreased in products with a complete assimilation to the expectations (G and T), particularly for the emotions characterized by a low degree of arousal (*boredom* and *indifference*).

These findings suggest that brand and packaging both contribute to increase liking and are potentially powerful elicitors of emotions, but only if this communication is perceived as coherent with the expectations. In cases of mismatch between what was communicated and what was experienced, informed liking confirmed the blind rating, but positive emotions tended to decrease. These results were in line with a previous study that showed that the packaging played a secondary role when compared to the sensory attributes of the product (Ng et al., 2013b). However, before generalizing these findings across all contexts, further studies on different product categories that systematically compare the impact both of packaging and of sensory characteristics of the product on emotions are needed. Preparatory semiotic analyses of the packages could be useful in such studies, as they allow mapping of the differences among packages, facilitating an understanding of the expectations that the packaging/branding build, as well as the meanings and emotions that drive consumer preferences.

5.3.4   Expectations and Emotions

Expectations play an important role in many emotions: surprise, curiosity, disappointment, and frustration are examples of emotions whose meaning is built on a mismatch between expected and actual experienced. In addition, there is a component of expectation in emotions elicited by an event and driven by an appraisal process that changes rapidly: emotions undergo constant modification, allowing rapid readjustment to changing circumstances and evaluations (Scherer, 2005; Scherer et al., 2006). Recently, Fontaine and Scherer (2013) proposed that a novelty dimension, in addition to valence (positive/negative), arousal (active/passive), and power (submissive/dominant), might be relevant in interpreting individual emotions (see also Fontaine and Veirman, 2013; Fontaine et al., 2007). This dimension was defined by appraisals of unpredictability and suddenness (experience of a novel event versus experiencing an emotion for a long time) and seems of particular interest for the studies on emotions elicited by commercial products. A new product can generate surprise that can be negatively or positively characterized (in term of valence) and that can vary in the long run, being replaced by a different emotion. Further studies on this dimension could help to develop predictive models based on emotions.

Fig. 5.7 is an attempt to summarize the relationships between the dynamics of liking (among the blind, expected, and informed conditions) and emotions, on the basis of the experience of the authors in the application of EmoSemio on both food and non-food products (Spinelli et al., 2014a,b,c, 2015).

A disconfirmation, a mismatch between the blind acceptability (B) and the expected acceptability (E), is defined as positive when expectations are lower than baseline product quality (B−E>0) and, conversely, as negative when expectations are greater than baseline product quality (B−E<0). In case of negative disconfirmation, an assimilation—liking moving to meet expectations—occurs if the informed liking is significantly higher than blind liking. The assimilation is complete if the informed liking is not significantly different from the expected liking, as informed liking scores are closer to the expected scores than the blind scores; inversely, consumers do not completely assimilate toward their expectations when the informed liking is significantly different from the expected liking. A contrast effect occurs when the discrepancy between expectancies and product performance is large (informed liking is significantly lower than blind liking) (Caporale et al., 2006; Siret and Issanchou, 2000; Cardello, 2007).

Complete assimilation of liking toward expectations is usually associated with an overall improvement of the emotional performance of the product: positive emotions increased in the case of complete assimilation toward the expectations, while negative emotions decreased. When there is a mismatch between expected liking evoked by packaging and blind liking, but an assimilation effect is not found, some positive emotions significantly decreased, demonstrating a worsening of the emotional performance of the products. This case is of particular interest because the informed liking did not significantly differ from the blind one here, but some positive emotions were significantly less intense in the informed condition than in the blind condition. If these data would be confirmed, emotions could provide additional information to interpret the product performance showing that there is trend to a worsening of the emotional performance, even if the liking rating still remains acceptable and unchanged.

From these first results it is apparent that the relationship between emotions and expectations is close and thus needs to be further investigated on different product categories. Collecting emotional responses in the expected condition as well could help in interpreting product performance.

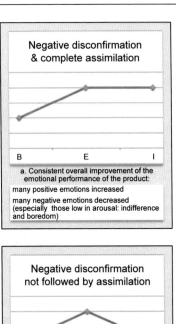

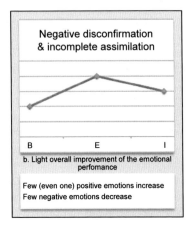

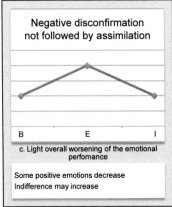

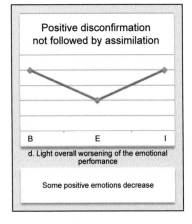

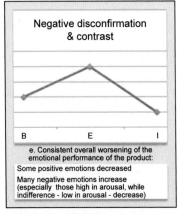

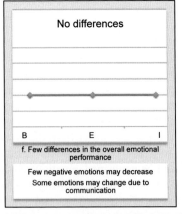

Figure 5.7 Relationships between the dynamics of liking passing from the blind (*B*), the expected (*E*), and the informed (*I*) condition and emotions. The line charts represent the liking measured in the three conditions and its interpretation in relationship to the theory of expectations (disconfirmation, assimilation, contrast). The captions summarize the emotions, comparing the B and the I condition (emotions were not collected in E condition).

5.4 Future Trends

During the past few years, research on emotions made great progress, particularly in the field of neuroscience. The understanding of the role of emotions in decision-making is growing, and recent findings are showing how these are implemented by processing in the brain. More research can be expected on the mechanism of reward and particularly on the emotional responses to food (Rolls, 2015).

The complexity of the research topic stresses the necessity to attack problems in an interdisciplinary way. From an applied point of view, the development of methodologies and particularly of the combination of different methods and approach is expected. The goal of the work should dictate the choice of the method, and the integration between different measures, explicit and implicit, will probably be more common in the future. Applied neuroscience techniques are appealing, but they are most useful as a complement, rather than a substitute, to existing methodology (Plassmann and Karmarkar, 2015). In addition, at present, there are still technical inconveniences in the application of these tools to study food product experience. The actual technical devices usually alter considerably the context, and this should be taken into account when the interest is measuring an emotional state. That can also prevent from conducting studies with real food and drink products, considering the technical problems related to the delivering of the product in a condition where any movement can compromise the measurement. On the other hand, in the last decade, self-report measures have been improved to a great extent and optimized for a specific application in measuring emotional food product experience, delivering fine-grained results. Studies on language in the field of psychology of emotions and linguistic can enrich theorizing and developing self-report measures in consumer studies, particularly in planning cross-cultural studies. Furthermore, the growing number of studies that utilize implicit methods suggests that they could also be valuable tools to investigate emotion associations, especially in combination with explicit methods.

The issue of context and how it can be implemented in the setting of the research is expected to be a hot topic, from the interest in the ecological setting to that in the virtual reality and virtual shelf testing.

The study of emotions in product experience is a promising field because it offers the opportunity to truly link the product to packaging and branding, passing from their sensory properties and the associations in the experiences of consumers. Further studies are needed to investigate more deeply the strict link between sensory characteristic of the products, hedonic, and emotional responses. Investigating the role of expectations, and particularly of emotional expectations, will give us a deeper insight into the product experience.

5.5 Sources of Further Information and Advice

The reference list at the end of the chapter is a starting point for sources of information on the topic of emotions and their measurement in a product context. Recently, three reviews were edited on the topic of emotions elicited by food, discussing methodological aspects and the current state of emotion research in product development (Jiang et al., 2014; Meiselman, 2015; Köster and Mojet, 2015). In addition, there is growing number of papers published every year on the topic on the Food Quality and Preference and Food Research International, which in 2015 dedicated a special issue to "Food, Emotions and Food Choice" (Varela and Ares, 2015).

References

Ares, G., Deliza, R., 2010. Studying the influence of package shape and colour on consumer expectations of milk desserts using word association and conjoint analysis. Food Quality and Preference 21, 930–937.

Ares, G., Piqueras-Fiszman, B., Varela, P., Marco, R.M., López, A.M., Fiszman, S., 2011. Food labels: do consumers perceive what semiotics want to convey? Food Quality and Preference 22, 689–698.

Bagozzi, R.P., Gopinath, M., Nyer, P.U., 1999. The role of emotions in marketing. Journal of the Academy of Marketing Science 27 (2), 184–206.

Balconi, M., Grippa, E., Vanutelli, M.E., 2015. Resting lateralized activity predicts the cortical response and appraisal of emotions: an fNIRS study. Social Cognitive and Affective Neuroscience 10 (12), pii: nsv041.

Becker, L., van Rompay, T.J.L., Schifferstein, H.N.J., Galetzka, M., 2011. Tough package, strong taste: the influence of packaging design on taste impressions and product evaluations. Food Quality and Preference 22, 17–23.

Belson, W.A., 1981. The Design and Understanding of Survey Questions. Gower, London.

Bennett, C.M., Baird, A.A., Miller, M.B., Wolford, G.L., 2009. Neural correlates of interspecies perspective taking in the post-mortem Atlantic Salmon: an argument for multiple comparisons correction. In: Poster Presented at the Society for Neuroscience Annual Meeting, Chicago, IL, USA.

Berridge, K.C., Winkielman, P., 2003. What is an unconscious emotion? (The case for unconscious "liking"). Cognition & Emotion 17 (2), 181–211.

Bhumiratana, N., Adhikari, K., Chambers IV, E., 2014. The development of an emotion lexicon for the coffee drinking experience. Food Research International 61, 83–92.

Bhumiratana, N., 2010. The Development of an Emotion Lexicon for the Coffee Drinking Experience (PhD Thesis). Kansas State University.

Bradley, M.M., Lang, P.J., 1994. Measuring emotion: the self-assessment manikin and the semantic differential. Journal of Behavior Therapy and Experimental Psychiatry 25 (1), 49–59.

Bradley, M.M., Lang, P.J., 2015. Memory, emotion, and pupil diameter: repetition of natural scenes. Psychophysiology 52 (9), 1186–1193.

Caporale, G., Monteleone, E., 2004. Influence of information about manufacturing process on beer acceptability. Food Quality and Preference 15, 271–278.

Caporale, G., Policastro, S., Carlucci, A., Monteleone, E., 2006. Consumer expectations for sensory properties in virgin olive oils. Food Quality and Preference 17, 116–125.

Cardello, A.V., Meiselman, H.L., Schutz, H.G., Craig, C., Given, Z., Lesher, L.L., et al., 2012. Measuring emotional responses to foods and food names using questionnaires. Food Quality and Preference 24, 243–250.

Cardello, A.V., 2003. Consumer concerns and expectations about novel food processing technologies: effects on product liking. Appetite 40 (3), 217–233.

Cardello, A.V., 2007. Measuring consumer expectations to improve food product development. In: MacFie, H.J.H. (Ed.), Consumer-led Food Product Development. Woodhead, Cambridge, UK, pp. 223–261.

Carrillo, E., Varela, P., Fiszman, S., 2012. Effects of food package information and sensory characteristics on the perception of healthiness and the acceptability of enriched biscuits. Food Research International 25, 110–115.

Carroll, D., Anastasiades, P., 1978. The behavioural significance of heart rate: the Laceys' hypothesis. Biological Psychology 7 (4), 249–275.

Chaya, C., Eaton, C., Hewson, L., Fernández Vázquez, R., Fernández-Ruiz, V., Smarte, K.A., Hort, J., 2015. Developing a reduced consumer-led lexicon to measure emotional response to beer. Food Quality and Preference 45, 100–112.

Chiao, J.Y., 2015. Current emotion research in cultural neuroscience. Emotion Review 7 (3), 280–293.

Chrea, C., Grandjean, D., Delplanque, S., Cayeux, I., LeCalve, B., Aymard, L., et al., 2009. Mapping the semantic space for the subjective experience of emotional responses to odors. Chemical Senses 34, 49–62.

Christoffersen, G.R., Schachtman, T.R., 2016. Electrophysiological CNS-processes related to associative learning in humans. Behavioural Brain Research 296, 211–232. http://dx.doi.org/10.1016/j.bbr.2015.09.011.

Clore, G.L., Ortony, A., Foss, M.A., 1987. The psychological foundations of the affective lexicon. Journal of Personality and Social Psychology 53, 751–766.

Coan, J.A., Allen, J.J.B., 2003. The state and trait nature of frontal EEG asymmetry in emotion." the asymmetrical brain. In: Hugdahl, K., Davidson, R.J. (Eds.), The Asymmetrical Brain (565–615). MIT Press, Cambridge.

Cohn, J.F., Zlochower, A.J., Lien, J., Kanade, T., 1999. Automated face analysis by feature point tracking has high concurrent validity with manual FACS coding. Psychophysiology 36 (1), 35–43.

Collinsworth, L.A., Lammert, A.M., Martinez, K.P., Leidheiser, M., Garza, J., Keener, M., Ashman, H., 2014. Development of a novel sensory method: image measurement of emotion and texture (IMET). Food Quality and Preference 38, 115–125.

Crisinel, A.S., Spence, C., 2009. Implicit association between basic tastes and pitch. Neuroscience Letters 464 (1), 39–42.

Dalenberg, J.R., Gutjar, S., ter Horst, G.J., de Graaf, K., Renken, R.J., Jager, G., 2014. Evoked emotions predict food choice. PLoS One 9 (12), e115388. http://dx.doi.org/10.1371/journal.pone.0115388.

Damasio, A.R., 1994. Descartes Error: Emotion Reason and the Human Brain. Avon, New York.

Damasio, A., 2003. Looking for Spinoza. Harcourt Inc., London.

Danner, L., Haindl, S., Joechl, M., Duerrschmid, K., 2014a. Facial expressions and autonomous nervous system responses elicited by tasting different juices. Food Research International 64, 81–90.

Danner, L., Sidorkina, L., Joechl, M., Duerrschmid, K., 2014b. Make a face! Implicit and explicit measurement of facial expressions elicited by orange juices using face reading technology. Food Quality Preference 32, 167–172.

Darwin, C., 1872. The Expression of the Emotions in Man and Animals. John Murray, London.

De Pelsmaeker, S., Schouten, J., Gellynck, X., 2013. The consumption of flavored milk among a children population. The influence of beliefs and the association of brands with emotions. Appetite 71, 279–286.

Desmet, P.M.A., Schifferstein, H.N.J., 2008. Sources of positive and negative emotions in food experience. Appetite 50 (2–3), 290–301.

De Wijk, R.A., Kooijman, V., Verhoeven, R.H.G., Holthuysen, N.T.E., de Graaf, C., 2012. Autonomic nervous system responses on and facial expressions to the sight, smell, and taste of liked and disliked foods. Food Quality and Preference 26, 196–203.

De Wijk, R.A., He, W., Mensink, M.G.J., Verhoeven, R.H.G., de Graaf, C., 2014. ANS responses and facial expressions differentiate between the taste of commercial breakfast drinks. PloS One 9 (4), e93823.

Delgado, C., Gomez-Rico, A., Guinard, J.-X., 2013. Evaluating bottles and labels versus tasting the oils blind: effects of packaging and labeling on consumer preferences, purchase intentions and expectations for extra virgin olive oil. Food Research International 54, 2112–2121.

Deliza, R., MacFie, H.J.H., 1996. The generation of sensory expectation by external cues and its effect on sensory perception and hedonic ratings: a review. Journal of Sensory Studies 11 (2), 103–128.

Delplanque, S., Grandjean, D., Chrea, C., Coppin, G., Aymard, L., Cayeux, I., et al., 2009. Sequential unfolding of novelty and pleasantness appraisals of odors: evidence from facial electromyography and autonomic reactions. Emotion 9, 316–328.

Delplanque, S., Chrea, C., Grandjean, D., Ferdenzi, C., Cayeux, I., Porcherot, C., et al., 2012. How to map the affective semantic space of scents. Cognition and Emotion 26 (5), 885–898.

Desmet, P.M.A., 2003. Measuring emotion: development and application of an instrument to measure emotional responses to products. In: Blythe, M.A., Monk, A.F., Overbeeke, K., Wright, P.C. (Eds.), Funology: From Usability to Enjoyment. Kluwer, Dordrecht, the Netherlands, pp. 111–123.

Di Monaco, R., Cavella, S., Di Marzo, S., Masi, P., 2004. The effect of expectations generated by brand name on the acceptability of dried semolina pasta. Food Quality and Preference 15, 429–437.

Eco, U., 1979. Lector in Fabula. Bompiani, Milano.

Eco, U., 1990. The Limits of Interpretation. Indiana U.P, Bloomington.

Ekman, P., 1989. The argument and evidence about universals in facial expressions of emotion. In: Wagner, H., Manstead, A. (Eds.), Handbook of Social Psychophysiology. Wiley, Chichester, England, pp. 143–164.

Ekman, P., Davidson, R.J., 1993. Voluntary smiling changes regional brain activity. Psychological Science 4 (5), 342–345.

Ekman, P., Friesen, W.V., 1975. Unmasking the Face. Prentice Hall.

Erk, S., Spitzer, M., Wunderlich, A.P., Galley, L., Walter, H., 2002. Cultural objects modulate reward circuitry. Neuroreport 13, 2499–2503.

Ferrarini, R., Carbognin, C., Casarotti, E.M., Nicolis, E., Nencini, A., Meneghini, A.M., 2010. The emotional response to wine consumption. Food Quality and Preference 21, 720–725.

Floch, J.M., 1990. Sémiotique, Marketing et Communication. PUF, Paris.

Floch, J.M., 1995. Identitées Visuelles. PUF, Paris.

Fontaine, J.J.R., Scherer, K.R., 2013. The global meaning structure of the emotion domain: investigating the complementarity of multiple perspectives on meaning. In: Fontaine, J.J.R., Scherer, K.R., Soriano, C. (Eds.), Components of Emotional Meaning. A Sourcebook. Oxford University Press, Oxford, pp. 106–125.

Fontaine, J.J.R., Scherer, K.R., Roesch, E.B., Ellsworth, P.C., 2007. The world of emotions is not two-dimensional. Psychological Science 18 (12), 1050–1057.

Fontaine, J.J.R., Veirman, E., 2013. The new NOVELTY dimension: method artifact or basic dimension in the cognitive structure of the emotion domain? In: Fontaine, J.J.R., Scherer, K.R., Soriano, C. (Eds.), Components of Emotional Meaning. A Sourcebook. Oxford University Press, Oxford, pp. 233–242.

Fontaine, J.R., Scherer, K.R., Soriano, C., 2013. In: Components of Emotional Meaning. A Sourcebook. Oxford University Press, Oxford.

Frijda, N.H., Scherer, K.R., 2009. Emotion definitions (psychological perspectives). In: Sander, D., Scherer, K.R. (Eds.), The Oxford Companion to Emotion and the Affective Sciences. Oxford University Press, Oxford and New York, pp. 142–144.

Frijda, N.H., 2008. The psychologists' point of view. In: Lewis, M., Haviland-Jones, J.M., Feldman Barrett, L. (Eds.), Handbook of Emotions, third ed. The Guilford Press, New York, pp. 68–87.

Gendron, M., Roberson, D., van der Vyver, J.M., Barrett, L.F., 2014. Perceptions of emotion from facial expressions are not culturally universal: evidence from a remote culture. Emotion 14 (2), 251–262.

Gibson, E.L., 2006. Emotional influences on food choice. Sensory, physiological and psychological pathways. Physiology Behavior 89 (1), 53–61.

Gmuer, A., Nuessli Guth, J., Runte, M., Siegrist, M., 2015. From emotion to language: application of a systematic, linguistic-based approach to design a food-associated emotion lexicon. Food Quality and Preference 40, 77–86.

Gouizi, K., Bereksi Reguig, F., Maaoui, C., 2011. Emotion recognition from physiological signals. Journal of Medical Engineering Technology 35 (6–7), 300–307.

Graham, D.J., Heidrick, C., Hodgin, K., 2015. Nutrition label viewing during a food-selection task: front-of-package labels vs nutrition facts labels. Journal of the Academy of Nutrition and Dietetics 115 (10). http://dx.doi.org/10.1016/j.jand.2015.02.019.

Greenwald, A.G., McGhee, D.E., Schwartz, J.L., 1998. Measuring individual differences in implicit cognition: the implicit association test. Journal of Personality and Social Psychology 74 (6), 1464–1480.

Greimas, A.J., Courtès, J., 1979. In: Sémiotique – Dictionnaire Raisonné de la Théorie du Langage. Hachette, Paris.

Guinard, J.X., Uotani, B., Schlich, P., 2001. Internal and external mapping of preferences for commercial lager beers: comparison of hedonic ratings by consumers blind versus with knowledge of brand and price. Food Quality and Preference 12, 243–255.

Gutjar, S., De Graaf, C., Kooijman, V., De Wijk, R.A., Nys, A., Ter Horst, G.J., Jager, G., 2014. The role of emotions in food choice and liking. Food Research International 76, 216–223.

Gutjar, S., Dalenberg, J.R., De Graaf, C., De Wijk, R.A., Palascha, A., Renken, R.J., Jager, G., 2015. What reported food-evoked emotions may add: a model to predict consumer food choice. Food Quality and Preference 45, 140–148.

He, W., Boesveldt, S., de Graaf, C., de Wijk, R.A., 2014. Dynamics of autonomic nervous system responses and facial expressions to odors. Frontiers in Psychology 13 (5), 110.

Hjortsjö, C.H., 1969. Man's Face and Mimic Language. Studen litteratur.

Huettel, S.A., Song, A.W., McCarthy, G., 2009. Functional magnetic resonance imaging. Massachusetts, second ed. Sinauer.

Jack, R.E., Blais, C., Scheepers, C., Schyns, P.G., Caldara, R., 2009. Cultural confusions show that facial expressions are not universal. Current Biology 19 (18), 1543–1548.

Jack, R.E., Garrod, O.G., Yu, H., Caldara, R., Schyns, P.G., 2012. Facial expressions of emotion are not culturally universal. Proceeding of the National Academy of Sciences of the United States of America 109 (19), 7241–7244.

Jackson, P.A., Kennedy, D.O., 2013. The application of near infrared spectroscopy in nutritional intervention studies. Frontiers in Human Neuroscience 7, 473.

Jaeger, S., 2006. Non-sensory factors in sensory science research. Food Quality and Preference 17, 132–144.

Jaeger, S.R., Cardello, A.V., Schutz, H.G., 2013. Emotion questionnaires: a consumer-centric perspective. Food Quality and Preference 30, 229–241.

Jager, G., Schlich, P., Tijssen, I., Yao, J., Visalli, M., de Graaf, K., et al., 2014. Temporal dominance of emotions: measuring dynamics of food-related emotions during consumption. Food Quality and Preference 37, 87–99.

Jiang, Y., King, J.M., Prinyawiwatkul, W., 2014. A review of measurement and relationships between food, eating behavior and emotion. Trends in Food Science Technology 36, 15–28.

Kagan, J., 2007. What Is Emotion?: History, Measures, and Meanings. Yale University Press, New Haven and London.

Kähkönen, P., Tuorila, H., 1996. How information enhances acceptability of a low-fat spread. Food Quality and Preference 7, 87–94.

Kähkönen, P., Tuorila, H., 1998. Effect of reduced-fat information on expected and actual hedonic and sensory rating of sausage. Appetite 30, 13–23.

Kahkonen, P., Tuorila, H., Rita, H., 1996. How information enhances acceptability of a low-fat spread. Food Quality and Preference 7, 87–94.

Kahkonen, P., Hakanpaa, P., Tuorila, H., 1999. The effect of information related to fat content and taste on consumer responses to a reduced-fat frankfurter and a reduced-fat chocolate bar. Journal of Sensory Studies 14 (1), 35–46.

Kahneman, D., 2003. A perspective on judgment and choice. American Psychologist 58 (9), 697–720.

Kenning, P.H., Plassmann, H., 2008. How neuroscience can inform consumer research. Neural Systems Rehabilitation Engineering, IEEE 16 (6), 532–538.

King, S.C., Meiselman, H.L., 2010. Development of a method to measure consumer emotions associated with foods. Food Quality and Preference 21, 168–177.

King, S.C., Meiselman, H.L., Carr, B.T., 2010. Measuring emotions associated with foods in consumer testing. Food Quality and Preference 21, 1114–1116.

King, S.C., Meiselman, H.L., Carr, B.T., 2013. Measuring emotions associated with foods: Important elements of questionnaire and test design. Food Quality and Preference 28, 8–16.

Köster, E.P., Mojet, J., 2007. Boredom and the reasons why some new food products fail. In: MacFie, H. (Ed.), Consumer-led and Food Product Development. Woodhead Publishing Limited, Cambridge, pp. 262–280.

Köster, E.P., Mojet, J., 2015. From mood to food and form food to mood: a psychological perspective on the measurement of food-related emotions in consumer research. Food Research International 76, 180–191.

Köster, E.P., 2009. Diversity in the determinants of food choice: a psychological perspective. Food Quality and Preference 20, 70–82.

Kreibig, S.D., 2010. Autonomic nervous system activity in emotion: a review. Biological Psychology 84 (3), 394–421.

Krishna, A., 2010. Sensory Marketing: Research on the Sensuality of Products. Routledge Chapman Hall, New York.

Krishna, A., 2012. An integrative review of sensory marketing: engaging the senses to affect perception, judgment and behaviour. Journal of Consumer Psychology 22 (3), 332–351.

Kuesten, C.L., Chopra, P., Bi, J., Meiselman, H.L., 2014. A global study using PANAS (PA and NA) scales to measure consumer emotions associated with aromas of phytonutrient supplements. Food Quality and Preference 33, 86–97.

Labbe, D., Pineau, N., Martin, N., 2013. Food expected naturalness: Impact of visual, tactile and auditory packaging material properties and role of perceptual interactions. Food Quality and Preference 27, 170–178.

Labbe, D., Ferrage, A., Rytz, A., Pace, J., Martin, N., 2015. Pleasantness, emotions and perceptions induced by coffee beverage experience depend on the consumption motivation (hedonic or utilitarian). Food Quality and Preference 44, 56–61.

Lange, C., Issanchou, S., Combris, P., 2000. Expected versus experienced quality: trade-off with price. Food Quality and Preference 11, 289–297.

Lange, C., Martin, C., Chabanet, C., Combris, P., Issanchou, S., 2002. Impact of the information provided to consumers on their willingness to pay for champagne: comparison with hedonic scores. Food Quality and Preference 13 (7–8), 597–608.

Laros, F.J.M., Steenkamp, J.-B.E.M., 2005. Emotions in consumer behavior: a hierarchical approach. Journal of Business Research 58, 1437–1445.

Larsen, J.T., Norris, C.J., Cacioppo, J.T., 2003. Effects of positive and negative affect on electromyographic activity over zygomaticus major and corrugator supercilii. Psychophysiology 40 (5), 776–785.

LeDoux, J.E., 1992. Emotion and the amygdala. In: Aggleton, J.P. (Ed.), Amygdala. Wiley-Liss, New York, pp. 339–351.

LeDoux, J.E., 1995. Emotions: clues from the brain. Annual Review of Psychology 46, 209–235.

LeDoux, J.E., 1996. The Emotional Brain: The Mysterious Underpinnings of Emotional Life. Simon and Schuster, New York.

LeDoux, J., 2008. Emotional coloration of consciousness: how feelings come about. In: Weiskrantz, L., Davies, M. (Eds.), Frontiers of Consciousness. Oxford University Press, Oxford, pp. 69–130.

LeDoux, J.E., 2011. Rethinking the emotional brain. Neuron 73, 653–676.

Leitch, K.A., Duncan, S.E., O'Keefe, S., Rudd, R., Gallagher, D.L., 2015. Characterizing consumer emotional response to sweeteners using an emotion terminology questionnaire and facial expression analysis. Food Research International 76, 283–292.

Lemercier, A., Guillot, G., Courcoux, P., Garrel, C., Baccino, T., Schlich, P., 2014. Pupillometry of taste: methodology guide - from acquisition to data processing and toolbox for MATLAB. The Quantitative Methods for Psychology 10 (2), 179–195.

Li, X.E., Jervis, S.M., Drake, M.A., 2015. Examining extrinsic factors that influence product acceptance: a review. Journal of Food Science 80, R901–R909.

Liao, L., Corsia, A., Lockshina, L., Chrysochou, P., 2012. Can packaging elements elicit consumers' emotional responses? In: 41st European Marketing Academy Conference. Lisbon, Portugal.

Liao, L.X., Corsia, A.M., Chrysochoua, P., Lockshina, L., 2015. Emotional responses towards food packaging: a joint application of self-report and physiological measures of emotion. Food Quality and Preference 42, 48–55.

Lindstrom, M., 2005. Brand Sense. Simon Schuster, New York.

Luu, S., Chau, T., 2009. Decoding subjective preference from single-trial near-infrared spectroscopy signals. Journal of Neural Engineering 6 (1), 016003.

Ma, Q., Wang, C., Wu, Y., Wang, X., 2014. Event-related potentials show taste and risk effects on food evaluation. Neuroreport 25 (10), 760–765.

Magnée, M.J., de Gelder, B., van Engeland, H., Kemner, C., 2007. Facial electromyographic responses to emotional information from faces and voices in individuals with pervasive developmental disorder. Journal of Child Psychology and Psychiatry, and Allied Disciplines 48 (11), 1122–1130.

Mai, R., Hoffman, S., Hopper, K., Schwarz, P., Rohm, H., 2015. The spirit is willing, but the flesh is weak: the moderating effect of implicit associations on healthy eating behaviors. Food Quality and Preference 39, 62–72.

Maison, D., Greenwald, A.G., Bruin, R.H., 2004. Predictive validity of the implicit association test in studies of brands, consumer attitudes, and behavior. Journal of Consumer Psychology 14 (4), 405–415.

Manzocco, L., Rumignani, A., Lagazio, C., 2013. Emotional response to fruit salads with different visual quality. Food Quality and Preference 28, 17–22.

Mauss, I.B., Robinson, M.D., 2009. Measures of emotions: a review. Cognition and Emotion 23 (2), 209–237.

McClure, S., Li, J., Tomlin, D., Cypert, K.S., Montague, L.M., Montague, P.R., 2004. Neural correlates of behavioral preference for culturally familiar drinks. Neuron 44 (2), 379–387.

Mehrabian, A., Russell, J.A., 1974. An Approach to Environmental Psychology, first ed. MIT Press, Cambridge, Mass.

Meiselman, H.L., 2015. A review of the current state of emotion research in product development. Food Research International 76, 192–199.

Mizutani, N., Dan, I., Kyutoku, Y., Tsuzuki, D., Clowney, L., Kusakabe, Y., et al., 2012. Package images modulate flavors in memory: Incidental learning of fruit juice flavors. Food Quality and Preference 24, 92–98.

Mojet, J., Dürrschmid, K., Danner, L., Jöchl, M., Heiniö, R.-L., Holthuysen, N., Köster, E., 2015. Are implicit emotion measurements evoked by food unrelated to liking? Food Research International 76, 224–232.

Mueller, S., Szolnoki, G., 2010. The relative influence of packaging, labelling, branding and sensory attributes on liking and purchase intent: consumers differ in their responsiveness. Food Quality and Preference 21, 774–783.

Mueller, S., Osidacz, P., Francis, I.L., Lockshin, L., 2010. Combining discrete choice and informed sensory testing to determine consumer response to extrinsic and intrinsic wine attributes. Food Quality and Preference 21, 741–754.

Mulligan, K., Scherer, K.R., 2012. Toward a working definition of emotion. Emotion Review 4, 345–357.

Murray, J.M., Delahunty, C.M., 2000a. Description of cheddar cheese packaging attributes using an agreed vocabulary. Journal of Sensory Studies 15 (2), 201–218.

Murray, J.M., Delahunty, C.M., 2000b. Mapping consumer preference for the sensory and packaging attributes of cheddar cheese. Food Quality and Preference 11, 419–435.

Nestrud, M.A., Meiselman, H.L., King, S.C., Lesher, L.L., Cardello, A., 2016. Development of EsSense25, a shorter version of the EsSense Profile®. Food Quality and Preference 48, 107–117.

Ng, M., Hort, J., 2015. Insight into measuring emotional response in sensory and consumer research. In: Delarue, J., Lawlor, B., Rogeaux, M. (Eds.), Rapid Sensory Profiling Techniques and Related Methods. Woodhead Publishing Limited, Cambridge, pp. 71–90.

Ng, M., Chaya, C., Hort, J., 2012. Liking sensory attributes obtained from QDA and TDS to liking and emotional response. In: Poster Presentation in 5th European Conference on Sensory and Consumer Research (Eurosense), Bern, Switzerland.

Ng, M., Chaya, C., Hort, J., 2013a. Beyond liking: comparing the measurement of emotional response using EsSense profile and consumer defined check -all-that-apply methodologies. Food Quality and Preference 28, 193–205.

Ng, M., Chaya, C., Hort, J., 2013b. The influence of sensory and packaging cues on both liking and emotional, abstract and functional conceptualisations. Food Quality and Preference 29, 146–156.

Nunez, P.L., Srinivasan, R., 2006. Electric Fields of the Brain: The Neurophysics of EEG, second ed. Oxford University Press, New York.

Ogarkowa, A., 2013. Folk concepts: lexicalization of emotional experiences across languages and cultures. In: Fontaine, J.J.R., Scherer, K.R., Soriano, C. (Eds.), Components of Emotional Meaning. A Sourcebook. Oxford University Press, Oxford, pp. 46–62.

Ortony, A., Clore, G.L., Foss, A.M., 1987. The referential structure of the affective lexicon. Cognitive Science 11, 341–364.

Parise, C.V., Spence, C., 2012. Assessing the associations between brand packaging and brand attributes using an indirect performance measure. Food Quality and Preference 27, 17–23.

Parkinson, B., 2005. Do facial movements express emotions or communicate motives? Personality and Social Psychology Review 9, 278–311.

Piqueras-Fiszman, B., Jaeger, S.R., 2014a. The impact of evoked consumption contexts and appropriateness on emotion responses. Food Quality and Preference 32, 277–288.

Piqueras-Fiszman, B., Jaeger, S.R., 2014b. Emotion responses under evoked consumption contexts: a focus on the consumers' frequency of product consumption and the stability of responses. Food Quality and Preference 35, 24–41.

Piqueras-Fiszman, B., Jaeger, S.R., 2014c. The impact of the means of context evocation on consumer's emotion associations towards eating occasions. Food Quality and Preference 37, 61–70.

Piqueras-Fiszman, B., Jaeger, S.R., 2015. The effect of product–context appropriateness on emotion associations in evoked eating occasions. Food Quality and Preference 40, 49–60.

Piqueras-Fiszman, B., Spence, C., 2011. Crossmodal correspondences in product packaging. Assessing color–flavor correspondences for potato chips (crisps). Appetite 57, 753–757.

Piqueras-Fiszman, B., Spence, C., 2012. The influence of the color of the cup on consumers' perception of a hot beverage. Journal of Sensory Studies 27, 324–331.

Piqueras-Fiszman, B., Spence, C., 2015. Sensory expectations based on product-extrinsic food cues: an interdisciplinary review of the empirical evidence and theoretical accounts. Food Quality and Preference 40, 165–179.

Piqueras-Fiszman, B., Ares, G., Varela, P., 2011. Semiotics and perception: do labels convey the same messages to older and younger consumers? Journal of Sensory Studies 26, 197–208.

Piqueras-Fiszman, B., Velasco, C., Spence, C., 2012. Exploring implicit and explicit crossmodal colour–flavour correspondences in product packaging. Food Quality and Preference 25, 148–155.

Piqueras-Fiszman, B., Velasco, C., Salgado-Montejo, A., Spence, C., 2013. Using combined eye tracking and word association in order to assess novel packaging solutions: a case study involving jam jars. Food Quality and Preference 28, 328–338.

Plassmann, H., Karmarkar, U.R., 2015. Consumer neuroscience. Revealing meaningful relationships between brain and consumer behavior. In: Norton, M.I., Rucker, D.D., Lamberton, C. (Eds.), The Cambridge Handbook of Consumer Psychology. Cambridge University Press, Cambridge.

Plassmann, H., Ramsøy, T.Z., Milosavljevic, M., 2012. Branding the brain: a critical review and outlook. Journal of Consumer Psychology 22, 18–36.

Plutchik, R., 1980. Emotion: A Psychoevolutionary Synthesis. Harper Row, New York.

Porcherot, C., Delplanque, S., Raviot-Derrien, S., Calve´, B.L., Chrea, C., Gaudreau, N., 2010. How do you feel when you smell this? Optimization of a verbal measurement of odor-elicited emotions. Food Quality and Preference 21, 938–947.

Porcherot, C., Delplanque, S., Planchais, A., Gaudreau, N., Accolla, R., Cayeux, I., 2012. Influence of food odorant names on the verbal measurement of emotions. Food Quality and Preference 23, 125–133.

Porcherot, C., Delplanque, S., Gaudreau, N., Cayeux, I., 2013. Seeing, smelling, feeling! Is there an influence of color on subjective affective responses to perfumed fabric softeners? Food Quality and Preference 27, 161–169.

Porcherot, C., Petit, E., Giboreau, A., Gaudreau, N., Cayeux, I., 2015. Measurement of self-reported feelings when an aperitif is consumed in an ecological setting. Food Quality and Preference 39, 277–284.

Quintana, D.S., Heathers, J.A., 2014. Considerations in the assessment of heart rate variability in biobehavioral research. Frontiers in Psychology 5, 805.

Rastier, F., 1997. Meaning and Textuality. University of Toronto Press, Toronto, (trans. Frank Collins and Paul Perron).

Richins, M.L., 1997. Measuring emotions in the consumption experience. Journal of Consumer Research 24 (2), 127–146.

Riezebos, H.J., 1994. Brand Added Value: Theory and Empirical Research about the Value of Brands and Consumers. Eburon, Delft.

Rolls, E.T., 2014. Emotion and Decision-making Explained. Oxford University Press, Oxford.

Rolls, E.T., 2015. Taste, olfactory, and food reward value processing in the brain. Progress in Neurobiology, 127/128. Oxford University Press, Oxford, pp. 64–90.

Sammler, D., Grigutsch, M., Fritz, T., Koelsch, S., 2007. Music and emotion: electrophysiological correlates of the processing of pleasant and unpleasant music. Psychophysiology 44 (2), 293–304.

Scherer, K., Dan, E.S., Flykt, A., 2006. What determines a feeling's position in affective space? A case for appraisal. Cognition and Emotion 20 (1), 92–113.

Scherer, K.R., 2005. What are emotions? and how can they be measured? Social Science Information 44 (4), 695–729.

Schifferstein, H.N.J., Desmet, P.M.A., 2010. Hedonic asymmetry in emotional responses to consumer products. Food Quality and Preference 21, 1100–1104.

Schifferstein, H.N.J., Spence, C., 2008. Multisensory product experience. In: Schifferstein, H.N.J., Hekkert, P. (Eds.), Product Experience. Elsevier, Amsterdam, pp. 133–161.

Schifferstein, H.N.J., Kole, A.P., Mojet, J., 1999. Asymmetry in the disconfirmation of expectations for natural yogurt. Appetite 32 (3), 307–329.

Schifferstein, H.N.J., Fenko, A., Desmet, P.M.A., Labbe, D., Martin, N., 2013. Influence of package design on the dynamics of multisensory and emotional food experience. Food Quality and Preference 27 (1), 18–25.

Schifferstein, H.N.J., 2010. From salad to bowl: the role of sensory analysis in product experience research. Food Quality and Preference 21, 1059–1067.

Schifferstein, H.N.J., 2015. Employing consumer research for creating new and engaging food experiences in a changing world. Current Opinion in Food Science 3, 27–32.

Siret, F., Issanchou, S., 2000. Traditional process: influence on sensory properties and on consumers' expectation and liking. Application to 'pâté de campagne'. Food Quality and Preference 11, 217–228.

Smith, V., Møgelvang-Hansen, P., Hyldig, G., 2010. Spin versus fair speak in food labelling: a matter of taste? Food Quality and Preference 21, 1016–1025.

Solomon, G.E., 2010. Niedermeyer's Electroencephalography: Basic Principles, Clinical Applications and Related Fields, sixth ed. Lippincott Williams Wilkins, Wolters Kluwer, Philadelphia.

Spinelli, S., Masi, C., Dinnella, C., Zoboli, G.P., Monteleone, E., 2014a. How does it make you feel? A new approach to measuring emotions in food product experience. Food Quality and Preference 37, 109–122.

Spinelli, S., Piccoli, B., Recchia, A., Zoboli, G.P., Monteleone, E., 2014b. When liking is not enough. Emotions as key for a better understanding of product performance. In: Poster Presented at Eurosense "A Sense of Life," Sixth European Conference on Sensory and Consumer Research, Copenhagen.

Spinelli, S., Piccoli, B., Zoboli, G.P., Monteleone, E., 2014c. How sensory and branding affect liking and emotions in delicate laundry detergents. In: Poster Presented at Eurosense "A Sense of Life," Sixth European Conference on Sensory and Consumer Research, Copenhagen.

Spinelli, S., Masi, C., Zoboli, G.P., Prescott, J., Monteleone, E., 2015. Emotional responses to branded and unbranded foods. Food Quality and Preference 42, 1–11.

Steiner, J.E., 1973. The gustofacial response: observation on normal and anencephalic newborn infants. Symposium on Oral Sensation and Perception 4, 254–278.

Thomson, D.M.H., Crocker, C., 2013. A data-driven classification of feelings. Food Quality and Preference 27, 137–152.

Thomson, D.M.H., Crocker, C., 2014. Development and evaluation of measurement tools for conceptual profiling of unbranded products. Food Quality and Preference 33, 1–13.

Thomson, D.M.H., Crocker, C., 2015. Application of conceptual profiling in brand, packaging and product development. Food Quality and Preference 40, 343–353.

Thomson, D.M.H., Crocker, C., Marketo, C.G., 2010. Linking sensory characteristics to emotions: an example using dark chocolate. Food Quality and Preference 21, 1117–1125.

Thomson, D.M.H., 2007. SensoEmotional optimization of food products and brands. In: MacFie, H. (Ed.), Consumer-led and Food Product Development. Woodhead Publishing Limited, Cambridge, pp. 281–303.

Thomson, D.M.H., 2015. Expedited procedures for conceptual profiling of brands, products and packaging. In: Delarue, J., Lawlor, B., Rogeaux, M. (Eds.), Rapid Sensory Profiling Techniques and Related Methods. Woodhead Publishing Limited, Cambridge, pp. 91–118.

Tuorila, H., Cardello, A.V., Lesher, L., 1994. Antecedents and consequences of expectations related to fat-free and regular-fat foods. Appetite 23, 247–263.

Tuorila, H., Andersson, A., Martikainen, A., Salovaara, H., 1998. Effect of product formula, information and consumer characteristics on the acceptance of a new snack food. Food Quality and Preference 9, 313–320.

Van Dam, Y.K., Van Trijp, H.C.M., 2007. Branding and labelling of food products. In: Frewer, L., Van Trijp, H.C.M. (Eds.), Understanding Consumers of Food Products. Woodhead Publishing Limited, Cambridge, pp. 153–180.

van Zyl, H., Meiselman, H.L., 2015. The roles of culture and language in designing emotion lists: comparing the same language in different English and Spanish speaking countries. Food Quality and Preference 41, 201–213.

Varela, P., Ares, G., 2015. Special issue title: "Food, emotions and food choice." Food Research International 76, 179.

Varela, P., Ares, G., Giménez, A., Gámbaro, A., 2010. Influence of brand information on consumers' expectations and liking of powdered drinks in central location tests. Food Quality and Preference 21, 873–880.

Villringer, A., Planck, J., Hock, C., Schleinkofer, L., Dirnagl, U., 1993. Near infrared spectroscopy (NIRS): a new tool to study hemodynamic changes during activation of brain function in human adults. Neuroscience Letters 154 (1–2), 101–104.

Violi, P., 1997. Significato Ed Esperienza. Milano: Bompiani (trans.: Meaning and Experience). Indiana University Press, Bloomington. 2001.

Watson, D., Clark, L.A., Tellegen, A., 1988. Development and validation of brief measures of positive and negative affect: the PANAS scales. Journal of Personality and Social Psychology 54, 1063–1070.

Williams, D.P., Cash, C., Rankin, C., Bernardi, A., Koenig, J., Thayer, J.F., 2015. Resting heart rate variability predicts self-reported difficulties in emotion regulation: a focus on different facets of emotion regulation. Frontiers in Psychology 6, 261.

Yan, X., Young, A., Andrews, T., 2015. Cultural similarities and differences in processing facial expressions of basic emotions. Journal of Vision 15 (12), 930.

Zeinstra, G.G., Koelen, M., Colindres, D., Kok, F.J., de Graaf, C., 2009. Facial expressions in school-aged children are a good indicator of "dislikes," but not of "likes." Food Quality and Preference 20, 620–624.

Neurosense and Packaging: Understanding Consumer Evaluations Using Implicit Technology

6

E. Fulcher, A. Dean, G. Trufil
Neurosense Ltd, Cheltenham Film Studios, Cheltenham, United Kingdom

6.1 Problems With the Self-Report Method in Market Research

The more "traditional" market research methodologies tend to be based on the concept of the *rational consumer*. In this view, it is often taken as fact that consumers know what they want and do what they can to get it. The rational consumer balances the costs and benefits of each option, selecting the one that gives them the greatest utility. Market research based on this view places the focus on *conscious decisions* or "rational" choices through interviews and surveys (which we will refer to as *self-report* or *explicit measures*). However, several problems with these assumptions and explicit measures in general have already been identified in previous research (eg, af Wåhlberg, 2009), and these include:

- Respondents may merely attempt to present themselves in favorable ways. The result is that the research is not an accurate account of how people feel and behave when making purchasing decisions. It is vital that researchers know how people actually behave rather than how they would like to be seen to behave.
- Respondents may only be consciously aware of global and generalized feelings, and specific memories (specific instances of their past behavior). The result is that the feelings expressed may lack detail and be based on the first memory that comes to mind, rather than a more accurate summary of one's behavior.
- Another result of the previous point is that respondents may overgeneralize a positive (or negative) feeling or preference toward a concept, so that all questions about that concept are rated or answered in the same way.
- Respondents may try to appear consistent in their responses. This results in a unidimensional model of behavior (which is consistent with a model of the rational consumer), but previous research suggests that attitudes often contain inconsistencies and contradictions.
- Respondents may try to be "kind" to the researcher. Most of the people who participate in market research receive a monetary incentive to do so, and may feel obliged to give brand-favorable responses.
- Respondents may deliberately try to hide feelings. This can happen when one is asking about certain behaviors that might be embarrassing, illegal, or socially less acceptable, but it can also happen if the respondent is concerned about the security of their data and how it might

Integrating the Packaging and Product Experience in Food and Beverages. http://dx.doi.org/10.1016/B978-0-08-100356-5.00006-1

be used. Respondents may want to hide their feelings for no obvious reason other than they just do not want to divulge the information.

- Respondents may simply make guesses about their true feelings. It has been argued that self-knowledge is often based on perception of one's own behavior, rather than by having a "privileged access" into our deeper feelings (Ryle, 1957). If we cannot introspect very accurately then a verbal response is most likely to be merely a rationalization of one's own behavior and not the genuine reason.
- Respondents purely wanting to get through a test as quickly as possible to receive a reward might use any strategy available to them for doing so. One of these is ticking boxes fairly randomly.
- Attitudes and feelings can also be very difficult to verbalize. A respondent may be aware of their feelings toward a brand but find it difficult to put into words. Some people are extremely good at this (eg, authors, copywriters, journalists, scriptwriters, etc., and people who are high on verbal intelligence) but most are not.

What is common in most of these problems is that responses are not being controlled, meaning that respondents can answer in any way they choose to, irrespective of how the researcher would like them to. If sufficient numbers of respondents behave in these ways then each problem is going to contribute a significant amount of noise to the data, resulting in weak discriminations between brands or design routes that are being tested. Putting all of these potential problems together, then it is clear that there is likely to be a large *truth gap* between what people say they will do and how they really behave in the marketplace. Drawing conclusions from this kind of data becomes hazardous.

While many explicit measures clearly have their place in market and consumer research (eg, when the respondent is genuinely *willing* to answer the question and *can access* relevant knowledge for an *accurate* answer), researchers have tried to look for alternative methods that promise to offer a deeper approach for measuring attitudes toward brands, products, and concepts, that might circumvent the sorts of problems just illustrated.

6.2 Products Are Evaluated Spontaneously by Consumers

Most of the things that we encounter in our everyday lives could be said to be "value charged," meaning that when they are encountered they provoke a mild to strong immediate emotional reaction (Betsh et al., 2001). When we encounter a value-charged stimulus, such as an object, person, face, event, location, photograph, and so on (possibly including the self and one's own behavior), some form of evaluation or appraisal of it is likely to have occurred before an overt emotional reaction. Some of the evaluations we make are intentional and done over an extended period of time, but most take place immediately (ie, rapidly) and involuntarily (ie, spontaneously without deliberate intention to do so). Such an immediate and unintentional response is referred to as an "automatic evaluation" or "cognitive appraisal" (Duckworth et al., 2002; Oatley and Johnson-Laird, 1996).

Brands, products, and packaging are no different; they too are value charged, being things that provoke automatic evaluations. Such evaluations can determine the specific brands a consumer would consider buying within a market category (known as their *consideration set*), and in our view also seem to be formed at the nonconscious level. It is highly likely that automatic consumer evaluations of brands originate from previous experiences with the brand (such as direct personal experience, being exposed to the experiences of others, perceptions of the brand's marketing, and social influences, such as fashion and social norms; in fact, all touchpoints). Such experiences shape our memory of brands and our attitudes toward them (Betsh et al., 2001). Yet because many evaluations will have been automatic, it may be difficult, if not impossible, to verbalize them and to say precisely how they have shaped our view of a particular brand. In short, we may not be consciously aware of the origins of our attitudes and feelings toward a brand, which can be influenced by other likes and dislikes too (Fulcher and Hammer, 2001).

Some have gone as far as to say that this highlights a general division of the mind: one division that makes these nonconscious automatic evaluations and appraisals, and another that makes conscious, deliberate, rational thoughts. The fast, nonconscious, probably irrational part of the mind has been referred to as *System 1*. This system can be likened to a parallel supercomputer that performs rapid computations on information coming in from the senses. This system helps us do those things we have learned to do automatically, like understanding human speech, seeing a coherent world through our eyes, and navigate a path through a crowded street without bumping into anyone. These are incredibly complex skills that we have acquired but have come to know how to execute almost effortlessly. It is in this system where evaluations and appraisals take place, and they do so on all of the information coming in from the senses. They go on nonconsciously because they need to be quick and efficient; they are involved in keeping us alive, safe, and well fed. They are also involved in social judgments, such as the perception of other people and social situations.

Everyday language reflects our collective awareness of nonconscious influences: some brands may be said to have "a certain something" or a "je ne sais quoi": qualities that are difficult to define or verbalize. Also, terms such as *a gut reaction, a hunch, a suspicion, an intuition, an instinctive knowing, a feeling in one's bones, a shiver down the spine, goose pimples, the butterflies,* and so on are part of common parlance and might be said to refer to the same concept. Also, in everyday terms, a common conceptualization is that it involves "a sense of knowing without knowing how one knows," (Epstein, 2010, p. 296). This is a concept frequently referred to in studies of the relationship between emotional and cognitive processing. Intuitions can be accompanied with physiological changes (the term "gut reaction" refers to a feeling in the stomach perceived as either pleasurable or disagreeable when faced with a choice and interpreted as an emotion or a feeling, Prinz, 2005), and specific patterns have been referred to in the psychological literature as *somatic markers* (Damasio et al., 1991). These come to be associated with specific situations and are said to provide implicit information about each option when faced with a choice. In addition, neurological responses such as activation in the amygdala can occur when making automatic evaluations (Öhman et al., 2007). In sum, it is generally accepted that the evaluation of things we encounter can occur nonconsciously, unintentionally, rapidly, and without our awareness that we have made such an evaluation, and also without an awareness of the experiential origins of the evaluation.

The slower, more conscious division, referred to as *System 2*, also makes evaluations and appraisals but requires time. Deliberating on the advantages and disadvantages of several options is the main business of System 2. Because human cognitive apparatus does not enable us to introspect with any degree of accuracy, we cannot examine System 1 processes in detail. Yet System 1 influences our decisions all the time and nonconsciously, and through processes that cognitive psychologists are only beginning to understand. When asked how we arrived at the decisions we make, we appeal to System 2 to analyze what it can do to generate plausible reasons, and these are the reasons we give to others. This idea, as we mentioned earlier, gave rise to the philosophical position that we may be nothing more than observers of our own behavior without any special privileged access into our inner thoughts (Ryle, 1957). One view, which we subscribe to, is that a decision a person makes (eg, choosing a particular brand) can often be based on System 1 processing (eg, their attention may have been grabbed by a can of *Mr. Sheen* and the claim on the pack "citrus fresh," as well as the visually pleasing lemon graphic without them being aware of it), yet the person may have the feelings that this choice of product was a conscious decision: a System 2 decision (it was well thought out and rational). In this case, the person has attempted to justify the purchase (by coming up with a plausible reason for it, such as "because it's a good polish" or even worse, "because it's nice"). The feeling of consciously owning the decision may have just been an illusion. These are the sorts of verbal responses market researchers are in danger of when they conduct self-report surveys; they may miss reasons for consumer decisions that are much closer to the reality. This is very important for brand managers to know because by using the right tool they get to know how their products are being perceived and ultimately how their designs will affect sales.

6.3 The Neuroscience Alternative

Given (1) the problems associated with explicit measures and (2) the fact that so many consumer decisions originate below the level of conscious awareness, researchers have turned toward neuroscience. This field has the promise of providing one or more objective methods that can perhaps take an accurate read of System 1 and reveal exactly how a person is feeling about their products and packaging.

Some of the neuro-research technologies applied to consumer science include brain scans (typically, functional magnetic resonance imaging), EEG (electroencephalographic recordings, which indicate patterns of gross brain activity), the so-called biometrics (heart-rate, electrodermal response—how much we sweat, breathing rate, and so on, which are all aspects of an emotional response), facial decoding (identifying emotional expression in the movement of facial muscles), eye tracking (monitoring the direction of eye gaze on an image), and tests of reaction time. These methods do not rely on verbal responses, being more objective measures of how a person is feeling. The problem with all such methods is interpreting what the "dials" are recording, which requires having a sufficient (and consistent) research body to turn to for guidance.

Notwithstanding problems of interpretation, each method appears to be able to provide a particular part of the overall picture. For example, if someone was asked to view a product, experience a service, or watch a TV advert, then biometric measures could indicate the general feelings the person has, a brain scan could show how strong the person feels toward it (among other things), eye tracking could reveal which aspects are visually attended to most often and in which order, and facial decoding could indicate the global emotional state the person is in. Each has its own strengths and weaknesses. For example, while brain scanning may have been thought to be the "purest" method, it can be too expensive for most marketing budgets, and in terms of the insights it can reveal, there are other less expensive methods that can deliver the goods much more efficiently (see Mauss and Robinson, 2009 for a review of measures of emotion).

One of the most popular of the so-called "neuromarketing" techniques is not really "neuro" at all, although it *is* based on assumptions about neural networks in the brain, which is that they are intrinsically associative. This is the implicit reaction-time or IRT. The idea is to provide respondents with two sorts of information: the first is information about a brand (such as the brand logo, a TV ad, a claim made about the brand, and so on) and the second is information about how someone might feel toward the brand (such as a word, phrase, or image), which we call an "attribute." The logic of the test is that if someone feels very positively toward a brand then they will make speeded reactions when seeing the brand alongside positive attributes. Conversely, they will have slowed responses when seeing the brand alongside negative attributes. This is a relatively simple technique (and an easy task to do for respondents), but it is one that can provide an almost forensic level of insights and is very difficult to fake (Chan and Sengupta, 2010). Furthermore, it is also a very powerful technique because it can measure those nonconscious, gut-reactions on which many purchases appear to be based. This is a technique that has its origins in Fazio et al. (1986) and has been extensively studied ever since trying to understand the cognitive component of human emotion. It is one of the less expensive neuro-research techniques, yet it can provide information at a granular level.

6.4   Implicit Reaction-Time Tests

In this section, we discuss the tasks involved in carrying out an IRT. In the subsequent section, we work through explanations of how IRTs elicit consumer evaluations from the perspective of associative memory. It should be noted that the case studies at the end of this chapter are based on *affective priming*, which is one type of IRT. The most commonly researched IRT is the Implicit Association Test or IAT (Greenwald et al., 1998), which is based on quickly identifying items in two categories (eg, one category might be the names of two competitor brands). By mixing up the categories over a series of blocks, respondents who have a strong attitude to one of the items or brands will be induced into making speeded responses on some trials and slowed responses on other trials. A statistical comparison between the reaction times in each block can yield which brand the respondent prefers.

6.4.1 Affective Priming

Affective priming (Fazio et al., 1995) is similar in that it is based on making categorical judgments, though it differs in very important ways. The original affective priming task is to detect target words as either one evaluative category (eg, *Happy*) or another (eg, *Sad*) as in the IAT by pressing a corresponding button or key on the keyboard. However, in this method, target words are preceded very briefly by "primes." These primes are either congruent with the target word (the prime is *Joy* when the target is *Happy*, or the prime is *Gloomy* when the target is *Sad*) or incongruent (the prime is *Gloomy* when the target is *Happy*, or the prime is *Joy* when the target is *Sad*). Congruent trials can be performed quicker and with fewer errors than they can be in incongruent trials.

In a branded version of affective priming, the targets might be two brands, such as *Pepsi* and *Tango*. Primes would be 20–40 attributes that one wishes to test against the two brands (such as *Quality, Trusted, Refreshing*, and so on). The logic is the same: if Pepsi is perceived as refreshing and Tango is less so in the eyes of a respondent, then *Pepsi* will be detected quicker than *Tango* when either are preceded by the attribute *Refreshing*. This would occur when *Refreshing* and *Pepsi* are congruent, and when *Refreshing* and *Tango* are incongruent or *less* congruent. Note that respondents do not make any evaluative judgments, all they are required to do is to detect the names of brands by pressing a key. The strength of association between an attribute and one of the brands is detected (or implied) by differences in reaction time to detect each brand when they are both preceded by the same attribute. A detailed example is given in Case Study 2, later in the chapter.

Results such as these have also been obtained when the primes are presented subliminally, that is, when the primes are presented so quickly that the individual cannot report seeing them (Greenwald et al., 1996). The method has been used to investigate a wide range of issues, such as racial prejudice (Olson and Fazio, 2003; Degner and Wentura, 2010; Bertram et al., 2013), social anxiety disorder (Lange et al., 2012), attitudes to light and dark skin tones (Stepanova and Strube, 2012), attitudes to foods with high fat content (Huggins, 2011), attitudes toward sex (Macapagal and Janssen, 2011), sexual preference (Snowden et al., 2008), perception of musical meaning (Steinbeis and Koelsch, 2011), and consumer evaluations of branding, products, packaging, and radio and TV advertising.

6.5   System 1 and Associative Memory Networks

As we stated earlier, people are continually making automatic evaluations, and these influence our behavior in all sorts of ways. In the discussion that follows, we offer one view of how this works mentally. It is a highly simplified model that we use to describe System 1 and hence how IRTs can measure consumer evaluations.

Our starting point is the assumption that when we see an object, an *internal representation* of it becomes active and we become conscious of seeing the object. For example, if we see a carton of milk, then an internal representation of the concept "carton of milk" becomes active. This is not a new idea. The Greek philosopher Aristotle considered the view that when two items or events closely occur together in time, links are formed between them in the mind (*the law of contiguity*). The strength of the link, he argued, is determined by the frequency with which the events occur together. These ideas were later built on by the physician David Hartley in the early 18th century, and by the philosopher JS Mill and physiologist Alexander Bain around the middle of the 19th century. Others since developed models of learning based on principles of association (eg, Pavlov, 1927; Skinner, 1957). More recently, the work in the area of artificial neural networks or connectionism has seen a resurgent interest in associative memory (Rumelhart and McClelland, 1986; Hinton and Anderson, 1989; Anderson and Bower, 2014).

Let us consider a simplified model of how knowledge might be represented mentally. We consider a network of knowledge as a collection of concepts. The idea is that a concept (representing a thing, place, event, person, and so on) may be represented by a "node" in the network and is connected to other nodes that it is semantically related to (see Fig. 6.1). In the brain, a node might be represented by a large collection of neurons and their firing patterns. When someone sees an object, such as a *fish* for

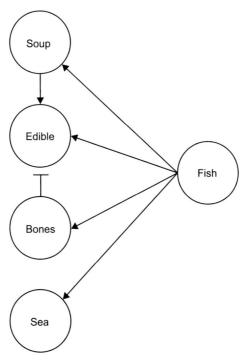

Figure 6.1 A simplified model of an associative network of knowledge. The *circled words* are nodes that represent concepts. When the concept *fish* becomes activated, it in turn activates concepts it is strongly associated (*linked*) with (those connect by *arrowed lines*). Note that the link between *bones* and *edible* is inhibitory (shown as a *T-connection*), reflecting the knowledge (at least in this network) that bones are not edible.

example, the node representing *fish* (and specific neurons in the brain) becomes active. If *fish* is strongly associated with other concepts in this network (eg, *soup*, *sea*, *bones*, and so on) then nodes representing those concepts will also become active (through a process called *spreading activation*). These new nodes can be activated to varying levels. Strongly activated concepts are the equivalent of remembered concepts (eg, when seeing a fish, one might be consciously reminded that it is edible), while moderately or weakly activated concepts do not enter conscious awareness (eg, activating the knowledge that it has bones is done nonconsciously).

A further development of this model is that spreading activation can occur between knowledge networks and emotion or evaluative networks and vice versa; (see Anderson and Bower, 2014; LeDoux, 2003; Rolls, 2005; Fulcher, 2001) and there is much evidence for this (see Subramaniam and Vinogradov, 2013; Binder and Desai, 2011). The properties of the model just described are similar to our current understanding of System 1. Concepts are linked through association, and their mental representations can be made active through these links and nonconsciously so. Active concepts are evaluated automatically and can influence a response. These ideas are integral in understanding implicit reaction-time methodology.

6.5.1 An Associative Memory Explanation

Now that we have mapped out a possible model of System 1, we can see how affective priming works. Suppose the attribute presented as a prime was *Refreshing* (continuing with the example in Section 6.4.1). This would activate the *Refreshing* concept node, which in turn, and through association (and spreading activation), would make the concept node *Pepsi* partially active. This would happen when there is a strong link between the two concepts nodes *Refreshing* and *Pepsi* in memory. So, when the word *Refreshing* is subsequently presented, it will not require much to make its concept node, *Pepsi*, active in memory, and this is because it is already partially active. This will result in a *fast* reaction time. Conversely, on trials where *Refreshing* is followed by *Tango*, the two concepts are weakly related, and so the concept node *Tango* will not get activated by the concept node *Refreshing*, so it will take longer for the concept node *Tango* to become active. This will result in a *slower* reaction time. Statistical analysis is done on each prime to determine whether it facilitated (speeded up) or inhibited (slowed down) reactions to both targets. The result is that the prime either facilitated one target but not the other, facilitated both targets, or inhibited both targets. This information can be used to infer how respondents associated each prime with each of the two targets (as we will see in the case studies, later).

6.6 Implicit Versus Explicit Measures: Validity Issues

In terms of the predictive validity of implicit versus explicit methods, there are numerous examples in the peer-reviewed literature demonstrating that in many circumstances implicit attitudes are better predictors of subsequent behavior than explicit responses provided at the same time. For example, Steinman and Karpinski (2009) found that implicit but not explicit attitudes toward the brand GAP predicted GAP patronage and buying intentions. Brunel et al. (2004) showed that implicit methods can detect attitudes about brands that explicit measures cannot (eg, how different races advocated different patterns of brand preferences implicitly but not explicitly).

Other research highlights the subtleties of attitudes that implicit methods can detect that are missed by explicit methods. For example, Priluck and Till (2010) found that explicit and implicit measures were both good at detecting attitudinal differences between brands when the difference was large or obvious, but only implicit methods could detect differences when they are less obvious. Other research shows that implicit methods in a consumer context are difficult to fake. For example, Chan and Sengupta (2010) found that while the claims of an advertisement were dismissed, implicit responses revealed that the ad had induced favorable attitudes to the brand.

A very convincing study of the benefits of assessing implicit attitudes is from Vianello et al. (2010) in which college students were given explicit and implicit measures of conscientiousness. Half were further told to imagine that they were being tested for their ideal job (good income, low effort, and so on) and half were not. Those with the job story had higher explicit measures of conscientiousness (showing that they could give biased answers), but both groups scored about the same on the implicit measure, which shows that the implicit measure was not easy to fake.

Neurosense also has a large body of unpublished data in which customer type is a strong predictor of the pattern of implicit responses, so for example, loyal customers of a shampoo provide a stronger and more positive pattern of implicit responses than do noncustomers. In addition, lapse customers have shown a more negative implicit response to the brand than noncustomers. In another study, fans of the horror movie genre showed a positive implicit response to a horror movie trailer, but everyone else showed a very strong aversive implicit response to the same. On occasion, we have found that some customers of a brand can reveal a more positive implicit profile of a competitor brand when the competitor is perceived more of a luxury than their regular brand (in terms of quality and price). Furthermore, using regression analysis, specific patterns of implicit responses have been shown to predict overall preference ratings of brands and estimates of costs. In addition, when we repeated a brand positioning study on a UK TV station 3 months apart, the pattern of responses was highly correlated.

6.7 Cognitive Psychology of Consumer Pack Perception

IRTs have been used for a broad range of market research issues and areas, such as brand positioning, brand equity, new product development, advertising, logo design and product naming, consumer segmentation, and recruitment. They are also used for packaging design, especially for comparing one or more new design routes with the existing pack and competitor packs.

Packaging is obviously not only about perception of the brand and whether or not a consumer will make a purchase, it is obviously very important for consolidation and protection of the contents and for providing legal information about the product. However, as products are nearly always evaluated automatically, effective pack design is critical for selling them; consumers often equate the package with the product itself. So for example, when one thinks of *Coca Cola*, the image of the bottle might come to mind before any recall of its taste. It is often said that the look and feel of the packaging is one of the company's last chances to convince consumers. It is also important for brand identification, especially for its target audience.

Some research suggests that up to 70% of a consumer's decisions about which products to buy are made in-store. When it comes to selection of specific brands, it has been estimated that 10% are regular switchers who may choose a different brand for the same product category each time, while 90% are loyal and tend to stay with their preferred brand. It is against this backdrop that brand differentiation (eg, shelf impact) becomes increasingly important. Factors such as the photograph or illustrations on a pack, its shape, the use of color and fonts are critical for communicating brand identity, for standing out among competitor brands, and appealing to the consumer. It has been said that the ideal package design is one the consumer would not want to throw away (Calver, 2007).

According to Rundh (2009), the final design of the packaging is influenced by different aspects of a business, such as manufacturing, distribution, and marketing, each with their own set of requirements. These place numerous constraints on the final design of the materials used, the shape of the pack, its size, its color, its texture, the photograph or illustration, other graphics, and the logo. Collectively, we might call these the products' assets. As the influences on pack design are continually changing, so a pack's assets require regular updating (Rundh, 2009). It is for these reasons that the pack's assets need to appeal to consumers. The collection of assets need to be (1) consistent or congruent (they need to have a consistent "tone of voice" or brand personality), (2) need to stand out visually (eg, shelf impact), and (3) need to appeal to the target audience (consistent with its brand positioning).

6.8 Case Studies

In the following section, we outline three case studies of projects undertaken by Neurosense for US and UK clients. The first one concerns a change to the packaging of a well-established brand and what the risks and opportunities might be; the second was carried out to compare the implicit and explicit measures of four similar brands from the same category. The third is a brief outline of a study we carried out for a large cosmetics company.

6.8.1 Case Study 1: A Popular Malt Drink

6.8.1.1 Background

Neurosense was commissioned by a UK drinks company to test several new packaging designs for one of their lines, which is a brand of milk-flavored drink made with a malt extract. Although tests were carried out on the original pack as well as a new front of pack (FOP), two new back of packs (BOPs), and the FOPs from three competitor brands, only the results of the original FOP against the new FOPs is reported here.

The key research question was how well the new design would deliver against the intended positioning of the brand and what its strengths and weaknesses are (eg, against competitor brands). It also focused on what areas might need to be retained and what areas might need to be improved in terms of pack design. Other areas of concern were whether there were any risks of the new design for existing loyal customers and whether there might be opportunities from noncustomers. These questions were approached by applying a BrainLink™ IRT to both FOPs to an online sample of current customers and a similar sample of competitor customers.

The current pack contains several assets, including the brand name in large bold letters across the front, a graphic to indicate physical exercise with words emphasizing the activity, a mug filled with the drink, and a list of the product benefits drawn in a dynamic way (ie, implying a pouring action). For the new pack, the physical exercise graphic and text is removed, the brand name is drawn so that it partly mimics a smiling face, and the mug with benefits are still shown but less dynamically than the current pack. The new pack has a brighter background color.

6.8.1.2   Method

The tests were carried out in 2014 at central locations in two different cities, chosen to represent two different types of demographics (middle and working classes), although tests themselves were taken online. The 560 respondents were mothers aged 18–65 years with children aged 3–10 years.

The main task of the test was to detect either an image of the main brand or one of the competitors (the target images). This was done by pressing one key as soon as they saw the main brand and another key as soon as they saw the competitor brand. After 20 practice trials at this task, the main trials began and required the same response, but this time preceding each target image was a word attribute or "prime." Word attributes reflected a broad range of feelings toward the product category. If, in the mind of the consumer, an attribute (such as *Loved*) is strongly associated with the main brand, then subsequent detection of the brand should be quicker than when the same attribute appears before the competitor brand. There were 46 attributes, repeated over several blocks of trials. Each respondent took more than one test, and the overall testing time was about 30 min (in general, online tests are much shorter, around 4–10 min).

As an additional test, respondents were shown all FOPs and BOPs and simply asked to examine the pack. While they were doing this, their eye movements were recorded through a standard webcam. The duration of eye gaze at each location was recorded.

6.8.1.3   Results

Across four BrainLink tests, reaction times were captured and collectively analyzed using principal component extraction with a varimax rotation (which is the most common method) to determine whether attributes would group into factors. Emerging factors would represent dimensions of similarly behaving attributes among the sample of consumers. Factors were interpreted as concepts deemed important for consumers when perceiving the brands, and inferred to be important when making decisions in the market category. Several factors emerged that reflect concepts related to family well-being, positive emotion, and quality of ingredients, personal relevance, and trust.

In the results for existing customers of the main brand, the current pack was strongly preferred over the new pack and on all factors, except for some emotional aspects. This is perhaps not surprising and one would expect customers to hold strong feelings toward a pack they are familiar with. However, this does not mean that new packs are instantly at a disadvantage in this kind of test due to less familiarity. As a company, we have commonly observed as many successful new packs as we have observed failures. This aside, some allowance for novelty would need to be considered in any evaluation of a change in packaging design; to that end, the emotional interest in the new pack is seen as one encouraging aspect.

For noncustomers, however, the new pack was preferred over the current pack on all bar one of these factors. The exception being the dimension of trust, which was more strongly associated with the existing pack. So while interest in the new pack from noncustomers conveys potential opportunities, one problem for this new packaging is a potential loss of trust from noncustomers, and that is a potential risk. Overall, the poor evaluation by existing customers and the poor score for trust for noncustomers indicates problems for the new packaging.

Turning to the results of the eye tracking component, there were some interesting findings. The first was that greater fixation time was given to the product benefits on the current versus new design, and a significant portion of fixation time was given to the active image with its accompanying text on the original pack. Greater fixation time was given to the brand name on the new pack when compared with the current pack.

6.8.1.4 Conclusions

The results of the implicit test and the eye tracking together imply that whereas the new brand name strongly features in the new design and potentially leads to positive emotional perception from its customers, the worse evaluation on other dimensions has most likely occurred due to the loss of the active image and the dynamic presentation of the cup and its benefits from the current design. Our recommendation to the client was therefore to retain the new way in which the product name is being displayed on the pack but to return to creative to determine how they could rebuild the active and dynamic features of the original pack.

6.8.2 Case Study 2: Bottled Wine Packaging

Neurosense in collaboration with Campden BRI conducted a proof-of-concept study to examine whether consumer evaluations measured by the implicit method were (1) sufficiently sensitive to capture consumer differences in four very similar products, (2) able to predict global explicit feelings (overall preferences and estimated product prices), and (3) whether this could be achieved better or worse than finer grained explicit evaluations of each product.

6.8.2.1 Method

Four moderately popular mid-priced brands of Sauvignon Blanc were chosen for this study King's Favour, Fairbourne, Vavasour, and Nautilus. These were all photographed from the same visual angle, and the labels, including the brand names, were clearly readable. Respondents were prescreened so that they must have consumed Sauvignon Blanc at least four times in the last 12 months. Respondents were presented with visual images of the bottles but did not sample any product.

BrainLink™ IRTs were applied to large samples of respondents ($n = 500$), each being tested on all four images. There were six IRTs in all so that each brand was tested against every other brand (see Fig. 6.2). Attributes were carefully selected from our database of attributes related to this market category and through desktop research.

6.8.2.2   Results

In terms of the implicit results, King's Favour was a very clear overall winner, topping in six of the seven dimensions. Fairbourne was most closely associated with sensory attributes and was the next most successful design, followed by Nautilus. Vavasour performed poorly overall, and it was likely that the name is a distinct factor in this, as "sour"

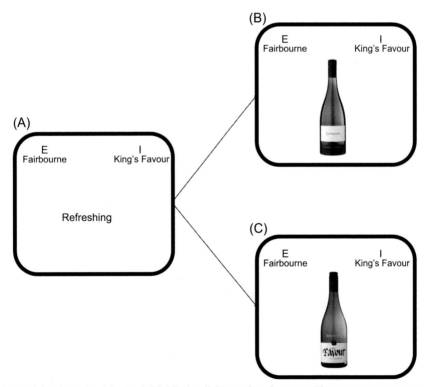

Figure 6.2 Diagram of the BrainLink™ implicit reaction-time test. There are two targets to look out for, Fairbourne and King's Favour bottles. As soon as they appear, the respondent presses the corresponding key on the keyboard (E for Fairbourne and I for King's Favour). First, an attribute such as *Refreshing* is presented (screen A). On some trials, the attribute is replaced with an image of Fairbourne (screen B), and the respondent has to press E, and on other trials the attribute is replaced by an image of King's Favour (screen C), and the respondent has to press I. The time difference between the two types of trials is used to infer which brand is more strongly associated with the attribute.

may have been perceived negatively as part of the brand name. Males favored King's Favour and Nautilus equally overall, while females preferred King's Favour followed by Fairbourne. King's Favour was also the clear favorite in the older age-group. In terms of explicit attitudes, there was little discrimination between the wines on a range of attributes with no statistical significance between them. There was, however, a direct relationship between the implicit profiles for each wine and the explicitly stated price estimation. The rank order of the implicit profiles for each wine matched the rank order of their price estimates from high to low. These data were entered into a regression analysis to model a specific pattern of attributes that collectively predicted the price of the wine.

6.8.2.3 Conclusion

Even though respondents did not sample the products, they clearly held strong implicit attitudes toward the four brands, and the implicit profiles were very distinct. The study showed that consumers make choices on product's visual assets, such as names, product shape designs, label colors, logos, and type fonts. The study also achieved its aim in showing that implicit preferences can be detected in similar products, and it can predict explicit attitudes or beliefs (such as the cost estimates of a product). It further showed that the implicit method was more discriminating than the explicit method in drawing out consumer perceptions and feelings (Figs. 6.3 and 6.4).

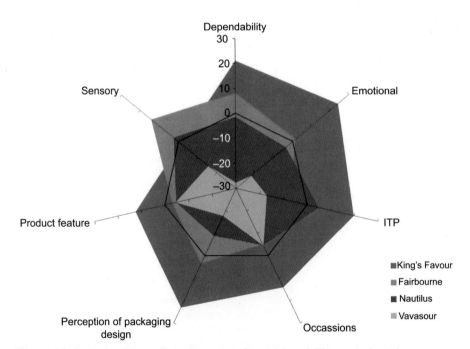

Figure 6.3 Analysis of the attribute dimensions for each brand. The *vertical axis* is a measure of the strength of association (and whether it is positive or negative) between the product and the attribute dimension. Data were amalgamated for each brand across the six tests. The larger the area, the more positive are the associations. *ITP*, intention to purchase.

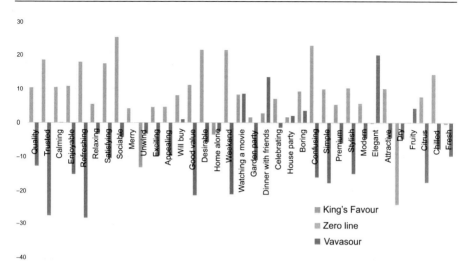

Figure 6.4 Chart showing the strength of association between each attribute used in the tests and two of the brands. The patterns of associations between the two brands are widely different.

6.8.3   Case Study 3: Packaging Assets Study for an American Beauty Products Company

6.8.3.1   Background

The client aimed to relaunch their range of beauty products by rebranding the packaging. They needed to identify the strongest and weakest visual assets of their packaging to inform new designs. The aims were therefore (1) to identify the key brand and packaging elements (colors, shapes, celebrities, pack patterns, typography, and icons) and a range of word attributes that are implicitly and intuitively associated with the brand and (2) to quantify the most iconic elements of the brand and points of differentiation with competitor brands. From this information, they would determine which valuable assets to retain and which design routes to follow.

6.8.3.2   Method

The study was carried out in two of the company's largest markets, the United States and China on 450 target brand users and competitor users. Respondents completed an online BrainLink™ IRT after completing a segmentation questionnaire and screener. Brand assets were used as primes in the test and consisted of 16 attribute words, five colors, two patterns, five different typographies, four product shapes, images of four celebrities, three selling lines or claims, and one brand logo, that are all used, or could be potentially used, by the brand on its packaging. Stimuli, especially the images of celebrities, differed to some extent between the two countries. Respondents took a number of BrainLink™ IRTs that identified (1) the strength of association between each asset and the brand, and (2) the strength of association between each asset and a competitor brand. These measures enabled us to analyze the data in terms of ownership of the assets by the brand (versus its main competitor) and how to best characterize the brand in terms of the descriptive assets (word attributes) and the visual assets.

6.8.3.3   Results

A number of key visual elements that are very strongly associated with the brand, and hence "owned" by the brand, were identified. The brand was also very strongly associated with several of its current selling lines and a number of new ones. Several typographies were also strongly associated with the brand. New assets that have great potential for the brand in enhancing its image and purchase intent were also uncovered. Several cultural differences were observed, such that the strength of some of the brand associations was different between China and the United States, and this applied to numerous colors, product symbols, and celebrities. Some of these were easily predictable (such as differences in the celebrities associated with the brand), but many were unexpected. Data cuts on age (above and below 35 years), brand usage, and consumer segment (using a three-dimensional model of consumers in this market category), provided further insights.

6.8.3.4   Conclusion

Through implicit research technology, Neurosense identified the strongest and weakest assets from the brand's current set. We also identified assets strongly associated with the product category in which the brand sites that were owned by neither brand, and hence which could be exploited. The business rebranded using the findings and recommendations by Neurosense. These designs are now in stores and on product shelves.

References

Anderson, J.R., Bower, G.H., 2014. Human Associative Memory. Updated Edition. Psychology Press.

Bertram, J., Schneider, J.F., Ewaiwi, A., 2013. Explicit and implicit measures of prejudicial attitudes toward muslims: on the relationship between explicit and implicit attitudes. In: Larsen, K.S., Vazov, G., Krumov, K., Schneider, J.F. (Eds.), Advances in International Psychology: Research Approaches and Personal Dispositions, Socialization Processes and Organizational Behavior. Kassell University Press, Kassel Germany, pp. 45–62.

Betsh, T., Plessner, H., Schwieren, C., Gütig, R., 2001. I like it but I don't know why: a value-account approach to implicit attitude formation. Personality and Social Psychology Bulletin 27, 242–253.

Binder, J.R., Desai, R.H., 2011. The neurobiology of semantic memory. Trends in Cognitive Science 15, 527–536.

Brunel, F.F., Tietjie, B., Greenwald, A.G., 2004. Is the implicit association test a valid and valuable measure of implicit consumer social cognition? Marketing 4, 385–404.

Calver, G., 2007. What Is Packaging Design? Rotovision, Hove.

Chan, E., Sengupta, J., 2010. Insincere flattery actually works: a dual attitudes perspective. Journal of Marketing Research 47, 122–133.

Damasio, A.R., Tranel, D., Damasio, H., 1991. Somatic markers and the guidance of behavior: theory and preliminary testing. In: Levin, H.S., Eisenberg, H.M., Benton, A.L. (Eds.), Frontal Lobe Function and Dysfunction. Oxford University Press, New York, pp. 217–229.

Degner, J., Wentura, D., 2010. Automatic prejudice in childhood and early adolescence. Journal of Personality and Social Psychology 98, 356–374.

Duckworth, K.L., Bargh, J.A., Garcia, M., Chaiken, S., 2002. The automatic evaluation of novel stimuli. Psychological Science 6, 515–519.

Epstein, S., 2010. Demystifying intuition: what it is, what it does, and how it does it. Psychological Inquiry 21, 295–312.

Fazio, R.H., Jackson, J.R., Dunton, B.C., Williams, C.J., 1995. Variability in automatic activation as an unobtrusive measure of racial attitudes: a bona fide pipeline. Journal of Personality and Social Psychology 69, 1013–1027.

Fazio, R.H., Sanbonmatsu, D.M., Powell, M.C., Kardes, F.R., 1986. On the automatic activation of attitudes. Journal of Personality and Social Psychology 50, 229–238.

Fulcher, E.P., 2001. Neurons with attitude: a connectionist account of evaluative learning. In: Moore, S., Oaksford, M. (Eds.), Emotional Cognition: From Brain to Behaviour. John Benjamins Publishing Company, Amsterdam.

Fulcher, E.P., Hammer, M., 2001. When all is revealed: a dissociation between evaluative learning and contingency awareness. Consciousness and Cognition 10, 524–549.

Greenwald, A., Draine, S.C., Abrams, R.L., 1996. Three cognitive markers of unconscious semantic activation. Science 273, 1699–1702.

Greenwald, A., McGhee, D.E., Schwartz, J.L., 1998. Measuring individual differences in implicit cognition: the implicit association test. Journal of Personality and Social Psychology 74, 1464–1480.

Hinton, G., Anderson, J., 1989. Parallel Models of Associative Memory. Erlbaum, Hillsdale, NJ.

Huggins, M., 2011. An Exploration of the Implicit Food Attitudes of People with Type-1 Diabetes Using Reaction-Time and Electrophysiological Measures (thesis). University of Birmingham, Birmingham.

Lange, W., Allart, E., Keijsers, G.P., Rinck, M., Becker, E.S., 2012. A neutral face is not neutral even if you have not seen it: social anxiety disorder and affective priming with facial expressions. Cognitive Behaviour Therapy 4, 108–118.

LeDoux, J., 2003. The emotion brain, fear and the amygdala. Cellular and Molecular Neurology 23, 4–5.

Macapagal, K.R., Janssen, E., 2011. The valence of sex: automatic affective associations in erotophilia and erotophobia. Personality and Individual Differences 51, 699–703.

Mauss, I.B., Robinson, M.D., 2009. Measures of emotion: a review. Cognition and Emotion 23, 209–237.

Öhman, A., Carlsson, K., Lundqvist, D., Invar, M., 2007. On the unconscious subcortical origin of human fear. Physiology of Behaviour 92, 180–185.

Oatley, K., Johnson-Laird, P.N., 1996. The Communicative Theory of Emotions: Empirical Tests, Mental Models, and Implications for Social Interaction. Lawrence Erlbaum Associates, Inc.

Olson, M.A., Fazio, R.H., 2003. Relations between implicit measures of prejudice: what are we measuring? Psychological Science 14, 636–639.

Pavlov, I.P., 1927. Conditioned Reflexes (Translated by W.H. Gantt.). International Press, New York.

Priluck, R., Till, B.D., 2010. Comparing a customer-based brand equity scale with the Implicit Association Test in examining consumer responses to brands. Brand Management 17, 413–428.

Prinz, J., 2005. Are emotions feelings? Journal of Consciousness Studies 12, 9–25.

Rolls, E.T., 2005. Emotion Explained. Oxford University Press, Oxford.

Rundh, B., 2009. Packaging design: creating competitive advantage with product packaging. British Food Journal 111, 988–1002.

Rumelhart, D.E., McClelland, J.L., 1986. Parallel distributed processing: explorations in the microstructure of cognition. In: Foundations, vol. 1. MIT Press, Cambridge, MA.

Ryle, G., 1957. The theory of meaning. In: Muirhead, J.H. (Ed.), British Philosophy in the Mid-Century. George Allen and Unwin, pp. 239–264.

Skinner, B.F., 1957. The experimental analysis of behaviour. American Scientist 45, 343–371.

Snowden, R.J., Wichter, J., Gray, N.S., 2008. Implicit and explicit measurements of sexual preference in gay and heterosexual men: a comparison of priming techniques and the implicit association task. Archives of Sexual Behavior 37, 558–565.

Steinbeis, N., Koelsch, S., 2011. Affective priming effects of musical sounds on the processing of word meaning. Journal of Cognitive Neuroscience 23, 604–621.

Steinman, R.B., Karpinski, A., 2009. The breadth-based adjective rating task as an indirect measure of consumer attitudes. Social Behavior and Personality: An International Journal 37, 173–174.

Stepanova, E.V., Strube, M.J., 2012. The role of skin color and facial physiognomy in racial categorization: moderation by implicit racial attitudes. Journal of Experimental Social Psychology 48, 867–878.

Subramaniam, K., Vinogradov, S., 2013. Improving the neural mechanisms of cognition through the pursuit of happiness. Frontiers in Heaven Neuroscience 7, 1–11.

Vianello, M., Robusto, E., Anselmi, P., 2010. Implicit conscientiousness predicts academic performance. Personality and Individual Differences 48, 452–457.

af Wåhlberg, A., 2009. Driver Behaviour and Accident Research Methodology. Ashgate Publishing Limited.

Further Reading

Fazio, R.H., 2007. Attitudes as objects-evaluations associations of varying strength. Social Cognition 25, 603–637.

Explicit Methods to Capture Consumers' Responses to Packaging

7

S. Thomas, M. Chambault
Campden BRI, Chipping Campden, United Kingdom

7.1 Introduction

Historically there has been a wealth of literature conveying the role and importance of packaging in terms of its functional properties and ability to communicate relevant product information, together with its influence on consumers' attention, perception, and purchase intention. More recently, there has been a growing interest surrounding the influence of the sensory characteristics of packaging on consumers' expectations and on consumers' subsequent food experiences. For brand owners and packaging design agencies, this presents a wonderful opportunity to differentiate and add value through effective packaging design. This chapter proposes a range of methods that packaging designers, consumer insights, and/or marketers could utilize to explore and quantify consumers' perceptions of the esthetic, functional, symbolic, structural, emotive, and informational components of packaging. The methods characterized in this chapter are predominantly applied at the front end of the development and assessment stages of packaging design. When used appropriately, these methods can generate consumer ideas and feedback, identify opportunities for innovation, and assist with documenting the business value of new concepts.

The chapter examines a diverse range of explicit methods that can be used with consumers to explore and measure their expectations, interactions, experiences, and satisfaction with product packaging. The chapter commences with a discussion of large-scale methods of quantitative assessment, followed by the use of small-scale focus groups and observational techniques. The discussion then progresses to more indirect methods that identify the relative importance of different packaging attributes, and to finish, holistic approaches that explore and categorize overall packaging or packaging design components are presented.

7.2 Large-Scale Quantitative Assessment of Consumers' Attitudes and Perceptions of Packaging Features

This first section will review some of the explicit, quantitative methods currently available to assess the product packaging with consumers and how the product packaging affects consumers' product choice and experience of the product.

Integrating the Packaging and Product Experience in Food and Beverages. http://dx.doi.org/10.1016/B978-0-08-100356-5.00007-3

Currently, products and their packaging are too often designed separately. It is also not common practice to assess the product packaging with the intended buyers, users, and/or consumers of the product. When it is the case, however, the product packaging is still too often assessed separately from the product and out of its context of purchase, usage, and/or consumption. Among the researchers who assess product packaging with consumers, some approach packaging as a set of individual elements such as sizes, shapes, colors, imagery, and typefaces: consumers are thought to evaluate each individual packaging element separately, which in turn affects their overall response/perception toward the packaging and/or the related product. Other researchers rather view the product packaging as a bundle of blended elements: their assumption is that consumers evaluate packaging in a holistic way rather than each individual element separately (Liao et al., 2015).

Besides brand and price, most of those researchers have focused their attention on the following extrinsic cues (ie, as opposed to the intrinsic characteristics pertaining to the product itself) or sources of information: names and sensory descriptions, information on food origin, growing process and production technologies, ingredients, nutrition- and health-related labels, claims, and emoticons (Vasiljevic et al., 2015), sustainable logos (Chambault, 2015), and pictorial cues. Those elements are assessed for the nature of the information they convey (eg, understanding and interpretation of the contents) as much as for how that information is conveyed (eg, impact of different sizes, shapes, fonts, and colors). For instance, Vasiljevic et al. (2015) showed that, with snack foods, emoticon labels yielded stronger effects on perceptions of taste and healthiness than color labels. Westerman et al. (2013) manipulated the graphic design of water and vodka bottles with respect to shape angularity, orientation, and left-right alignment and showed how congruent design elements (graphics and shape) may be advantageous in terms of purchase intent.

In general, most assessments are conducted at a test venue or online, but a few recent applications have been reported in more natural settings: in supermarket test halls (Becker et al., 2011) and a cafeteria (Gutjar, 2014). Gutjar (2014) showed that food choice in a simulated cafeteria setting is guided by extrinsic (packaging) as well as intrinsic (sensory) properties and that both can act as a cue for product appropriateness given a specific consumption context. When the packaging under evaluation cannot be physically shown to consumers, that is, in online studies or for practical reasons, good quality physical or digital images (Becker et al., 2011; Westerman et al., 2013) of the packaging are shown instead.

The concept of "multisensory experience" is still relatively new: even when consumers have in front of them the whole packaging to assess, most packaging assessments focus on visual cues (eg, shape and overall design) but rarely on odor, sound, and tactile/texture properties. Labbe et al. (2013) recently explored the perceptual interactions between and relative contribution of visual, auditory, and tactile stimulations coming from the packaging of dehydrated soup: consumers rated their expectation for the naturalness of the food based on its packaging under two test conditions: a "unimodal" condition, where consumers assessed separately the visual, tactile, and auditory stimuli from the materials, and a "trimodal" condition, where consumers simultaneously assessed the visual, tactile, and auditory stimuli from the materials.

Most assessments tend to focus on the product packaging only and/or the interaction between the product and its packaging, essentially through the measurement of not only consumers' product overall liking and purchase intent, but also perceived sensory characteristics and emotions/feelings before, during, and/or after consumption. More recently, overall satisfaction, as a new concept for measuring the integrated product-packaging experience, has also been introduced for food items (Burgess, 2015).

To ensure that consumers can easily recognize the product's main benefits (defined in the concept) and applications conveyed mostly by its packaging, although not widely reported in the literature, the product concept (printed, pictorial, or a mocked-up representation and description of the proposed product) would ideally be assessed along both proposed product and packaging at the early stages of the product life to ensure the so-called fundamental "product-concept fit." Prior to tasting the product, respondents would be presented with one or several concepts and asked to assess them for a series of attributes and/or statements and overall liking (typically on 9-point and 5-point JAR scales). Then respondents would be asked to assess the product packaging and the product itself (or several candidates)—separately then together—for how well they meet the expectations created by the concept(s) and using the same attributes/statements as those used to assess the concept(s) in addition to overall liking, purchase intent, and key sensory and more subjective attributes (eg, healthiness). Key concept performance measures could also be measured: uniqueness, relevance (satisfying a recognized need), credibility (belief that the product can really deliver the proposed benefits), and expected purchase frequency. Level of fit (not just "product-concept" but "packaging-product-concept") would then be estimated with correlation-based analyses. Preference mapping could also be used to determine the main drivers of consumer liking and penalty analysis to gain an understanding of the product-packaging attributes that most affect liking and purchase intent of the product-packaging entity. Carr et al. (2001) showed that fit-to-concept ratings related more to image (ie, subjective) attributes than to product description (ie, more objective) attributes.

The assessment of the impact of the packaging on the product typically involves assessing the product under two or three conditions (Mueller and Szolnoki, 2010; Chen, 2014): under blind (only the product, unbranded, is assessed), packaging/expected (only the packaging is assessed), and informed (the product is assessed branded, ie, in its packaging) conditions. In a three-stage assessment, Chen (2014) demonstrated that consumers' liking of wine products was significantly higher under the informed than under the blind condition for most of the products and that different levels of consumer emotional responses were obtained from the products under the informed but not under the blind condition. This emphasized the opportunity to utilize the emotional associations between packaging and product for defining a brand identity and consumer proposition. In this respect, emotional responses can be measured using either self-report (as in Chen, 2014) or physiological measures, namely facial electromyography and skin conductance. At present, a large stream of research uses self-report approaches, especially for assessing discrete emotions or subjective feelings (eg, surprise) utilizing instruments like the EsSense profile, the profile of mood states, the manual for the multiple affect adjective check (MAACL), and the Self-Assessment Manikin scale. Liao et al. (2015)

used the latter, consisting of 9-point scales with pictorial facial expressions, to measure consumers' emotional reactions to various packaging designs in terms of valence, arousal, and dominance. However, given that emotion can be spontaneous, fleeting, subconscious, and hard to be verbalized, self-report measures are often criticized (Liao et al., 2015).

To understand the cognitive process behind consumers' perception and distinction between product categories based on their whole packaging, some authors have used categorization tasks. Arboleda Ana and Arce-Lopera (2015) used such a task to assess how consumers visually recognize and distinguish between seven soft drink categories based on their bottle silhouette/shape by asking consumers to assign 52 bottle silhouettes to one of the proposed categories (eg, fruit juice) via radio buttons. Correlating the collected consumer data to analytical data (image analysis of the bottle physical characteristics) and using real bottles silhouettes, they found that the visual characteristics consumers use the most to differentiate between product categories were lid width and bottle shape (body kurtosis). Using artificially created bottle silhouettes, interestingly, they also found that, when the shape of the bottles was modified, consumers were not always capable of recognizing their corresponding product category.

However, attribute-based methods remain the most frequently employed. As opposed to holistic approaches (see dedicated section later in this chapter), they typically use tasks which involve rating the intensity of attributes on, for example, 9-point category scales (eg, liking) or 5-point JAR scales (eg, perceived sensory attributes), scoring or rating the packaging or product strength of association with shape, color, image, or description (Ngo et al., 2012), rating statements (eg, on product and packaging related expectation and usage) for importance or agreement using 5-, 7-, or 9-point category scales, selecting options (eg, sensory, functional properties and emotion terms) using CATA or MaxDiff scales (see dedicated section later in this chapter) or ranking (eg, products, packaging options, or statements). Overall, quantitative methods to assess food packaging are far more limited than those used to assess the product itself; for example, no temporal methods are utilized and objective characterization and quantification with a sensory panel is limited. Naming only a few examples, mostly based on visual cues and in no particular order, attribute-based methods aim to capture consumers':

- packaging perceived physical characteristics (eg, weight);
- skepticism toward specific types of on-pack information (Fenko et al., 2016);
- product and/or packaging evocations (ie, associations with a time, location, object, or person based on previous experiences), functional properties, convenience aspects (Westerman et al., 2013), and overall liking;
- product choice (Gutjar, 2014), preference, purchase intent, level of recommendation to relatives (Delgado et al., 2013), but also desire to consume (Delgado et al., 2013), intended usage (ie, context of use), perceived quality, naturalness (Labbe et al., 2013), refreshing sensation (Westerman et al., 2013), healthiness (Vadiveloo et al., 2013), tastiness (Vasiljevic et al., 2015; Westerman et al., 2013), nutritional expectations (Delgado et al., 2013), expected density (Tu et al., 2015) or calorific content (Vasiljevic et al., 2015), expected and perceived satiety (Vadiveloo et al., 2013; Tu et al., 2015), estimated amount consumed and

subsequent consumption of other foods (Vadiveloo et al., 2013), and price acceptance or expectation (Becker et al., 2011; Delgado et al., 2013);

• overall product-packaging satisfaction (Burgess, 2015) in addition to product packaging–derived emotions and feeling pre-, during, and/or post-consumption (Chen, 2014; Liao et al., 2015) as mentioned earlier.

Piqueras-Fiszman et al. (2012; cited in Tu et al., 2015) showed that perceived physical properties (weight) of container can exert significant influence on expected satiety and product density; Vadiveloo et al. (2013) assessed how labeling a pasta salad as "healthy" or "hearty" influenced self-reported satiety, consumption volume, and subsequent consumption of another food and showed that subtle food descriptions influenced post-intake experience of satiety.

The data collected with consumers is often interpreted in light of their attitudes and behaviors, in addition to their demographics as was traditionally the case. In terms of consumer demographics, as with product assessment, most packaging studies mostly involve adult participants. Only a few packaging studies have been conducted with children: Schouteten et al. (2015) conducted a CLT study with 10- to14-year-old children on Speculoos biscuits assessed in blind (unbranded products), expected (product logos), and informed (products and their logo) conditions using a successful combination of 9-point hedonic and CATA scales for the evaluation of perceived sensory characteristics and emotions.

The independent assessment of the product and its packaging does not lead to the same outcomes as their combined assessment, the interaction between the two, which leads to specific and so key product expectations, being neglected. Context also cannot be neglected; for example, product expectations derived from the packaging are dependent on the product's intended usage. Researchers are increasingly recognizing the importance of placing respondents in the right frame of mind when evaluating products. In a recent study (Burgess, 2015) on the perception of packaging design for chilled greens teas and its contribution to product satisfaction, a choice task was used at the start of the assessment to place participants into the right context: that of their usual local supermarket where they have to choose one product to buy among many others. In online studies, a couple of sentences and images are sometimes used to recall particular contexts of product purchase, use, or consumption.

To conclude, the choice of a method depends obviously on research objectives and resources, that is, budget and time frame, but also on the product-packaging type, the number of products/packaging to assess, and the type and number of consumers targeted. Online studies can offer a greater number and diversity of consumers compared to studies conducted at a test venue, but they do not come without their own disadvantages. Nowadays the food and drink industry appeals more and more to the consumers' experience expectations upon using the product than the consumers' sensory expectations: with origins from the field of experiential marketing, the focus is on the types of messages that can appeal to the emotional side of the customers, users, and/or consumers of the product-packaging entity through offering the creation of new experiences. Direct, explicit quantitative approaches are currently the most practical to implement with a large pool of consumers for most research objectives. Various alternatives to direct questioning of consumers have been proposed, including the analysis of facial expressions and the application of neuroscience tools to measure brain activity. However, these methods have not been developed to the point where they can be considered practical, proven research tools. Thomson and Crocker (2014, 2015) have considered the role of concepts, sometimes referred to as implicit associations or conceptual associations, as an alternative to understand what motivates consumers' choice behaviors. For them, consumers are motivated to buy products that ultimately deliver reward (the fundamental goal that drives human behavior) most effectively. As certain aspects of reward may occur below the level of conscious awareness, Thomson et al. argued that research involving direct questioning of consumers can be inappropriate, and that it is widely recognized that formal measurement of liking or purchase intent obtained in such a way can fail to predict consumption behavior.

7.3   Small-Scale Exploration of Consumers' Attitudes Toward Packaging Concepts and Prototypes: Focus Groups, Enabling, and Projective Techniques

This section will focus on the methods classically used in consumer and market research to enable consumers to generate vocabulary, visualize, and/or rationalize decision-making situations, as well as to explore and evaluate their thoughts and feelings in relation to concepts and/or prototypes. It will examine the projective and enabling techniques of word association and sentence completion, as well as the method of focus groups.

According to Gordon and Langmaid (1988) and Donoghue (2000), projective techniques consist of a situation or stimulus that encourages a person to project part of himself or herself or an idea system onto an external object or to bring it into the interview itself. Projective techniques help a researcher enter the private worlds of participants to uncover their inner perspectives, and reveal unconscious feelings and attitudes without them being aware that they are doing so, in a way they feel comfortable with. Enabling techniques create a situation that enables an individual's awareness to be focused on issues with which they are concerned. Enabling techniques are often considered simpler to use, they help participants to discuss issues, products, or packaging by using objects or scenarios in their minds. Will et al. (1996, p. 38) state that *"all projective techniques may be enabling techniques, but not all enabling techniques necessarily involve projection."* The key point about projective and enabling techniques is that respondents are asked to interpret the behavior of others, rather than directly asking them to report their beliefs and feelings. In interpreting the behavior of others, the respondents are indirectly projecting their own beliefs and feelings into the situation (Will et al., 1996; Kinnear and Taylor, 1991).

Word association is a projective technique which is increasingly being used to investigate consumers' perceptions of products (Ares and Deliza, 2010a; Guerrero et al., 2010 and Roininen et al., 2006). Association techniques require a participant to respond to the presentation of a stimulus with the first thing or things that come to mind. The stimulus could be in the form of an image, a statement or phrase, materials, or simply a question, such as *"What is the first thing you think of when I say 'X' Brand?"* In relation to each stimulus, participants are asked to say or write down the first words that come to mind. Word association is useful for eliciting information about consumer vocabulary associated with a brand, product, or packaging design.

Ares and Deliza (2010a) successfully applied word association to the study of package shape and color on consumer expectations of milk desserts. They asked participants to imagine the package and label of a milk dessert and to write down all the words, descriptions, associations, thoughts, or feelings that came to mind. They concluded that word association was a useful, efficient, and quick method to determine the most salient package and label features in consumers' minds that might influence their perception of milk desserts. However, they suggest that further research is necessary to understand the relationship between the frequency and order with which the terms are mentioned. In this instance, the frequency of mention was taken as an indicator of relevancy for each term. Comparative studies using methods such as conjoint analysis and word association could be used to investigate whether the most mentioned features are really the most important attributes influencing consumer purchase decisions.

In a different study by Ares and Deliza (2010b), which utilized conjoint analysis and word association to examine consumer expectations of milk desserts packaging shape and color, they found that consumers' responses via the word association were predominantly concerned with the sensory characteristics of the desserts, mainly flavor and texture attributes. Interestingly, the consumer associations were highly significantly different for the six evaluated packages, suggesting that package shape and color significantly affected consumers' sensory and hedonic expectations of the desserts within the packages. With regard to word association as a method for examining consumers' perceptions of packaging, Ares and Deliza (2010b) concluded that word association is able to capture consumers' spontaneous responses to stimuli, and therefore it could be used within evoked contexts to provide powerful insights to consumers' associations with food packaging, particularly at the pre-purchase stage of product selection.

Sentence completion is an enabling technique whereby participants are required to complete sentences, stories, arguments, or conversations. This type of technique is useful for eliciting underlying feelings and attitudes toward a particular product or brand. The technique can be extended to become a construction technique, whereby a participant is asked projective style questions in relation to images or bubble drawings. This type of projective questioning requires participants to give their opinions of other people's actions, feelings, or attitudes. Bubble drawings give participants the opportunity to complete or fill in the speech bubbles in a drawing or image. The idea is that they project their own opinions on to the thoughts or words of the person in the drawing or picture/image. The product, packaging designs, or labeling information can easily be incorporated into the image or drawing and made a focus of the stimulus on which participants need to respond.

Masson et al. (2016) undertook an interesting methodological study to assess participants' perceptions of coffee cups. The study utilized six qualitative methods, three of which were word association, sentence completion, and focus groups. Prior to the assigned task, participants were presented with eight coffee cups of different shapes, materials, and colors, and asked to look and handle them. With regard to the word association task, participants were asked about all the words that came into their minds for each of the eight cups. For the sentence completion task, participants were asked to complete six incomplete predefined sentences relating to the experience of drinking coffee. This was repeated for each of the eight cups. Meanwhile, the focus group explored the universe of coffee drinking and participants' perceptions of each cup. The results demonstrated major differences in the quantity of terms (subjective dimensions, attributes, and evocations) elicited by each method. Word association, sentence completion, and focus groups were among those methods that generated the largest number of terms. In particular, the study concluded that word association and sentence completion techniques were most effective in identifying the subjective dimensions, for example, colors, shapes, materials, patterns, and symbolic aspects of the coffee cups.

Typically projective and enabling techniques are conducted within the context of one-to-one interviews (Masson et al., 2016; Guerrero et al., 2010; Ares and Deliza, 2010a; Roininen et al., 2006), although, with an accomplished moderator, there is an opportunity to use these techniques within a focus group setting (Will et al., 1996) providing there is sufficient duration and the group has reached the "performing" stage of group development (Tuckman, 1965).

Focus groups consist of six to eight participants that have been recruited and screened for the specific purpose of discussing a particular issue, event, product, or packaging. Focus groups vary in size, style, and duration, although they are typically used at the front end of any innovation and development activity. For example, participants might be brought together as a group to exchange their experiences of using particular types of packaging, capture responses to existing and/or prototype packaging, generate ideas for new components of design, or identify and discuss their frustrations, as well as the benefits and/or environmental concerns of excessive or non-recyclable packaging. When used at the front end of product and packaging development, focus groups can provide richer detail on important and often unique information about consumers' needs, motivation, and values, for example, to identify the product features that have the most influence on the subjective requirements of consumers. By showing consumers unbranded packaging from different countries and/or cultures, it can push consumers to think beyond the scope of current offerings and facilitate the creation of new designs and/or use of alternative materials for packaging of food and drink products.

Depending on the length and nature of the focus group, participants can undertake a range of different activities, such as discussion, brainstorming, word association, sentence completion, tactile assessment of the packaging formats, and/or product taste evaluations in informed or blind conditions (with or without packaging). Tactile assessments, in the sense of participants interacting with existing, prototype, or packaging materials, is an opportunity to gain valuable insights into consumers' perceptions of the functionality, storage, and sensory properties (touch, smell, and sound). As Young (2004, p. 70) suggests *more can be learned from watching 20 people actually hold, use and discuss a new packaging system than from getting quick ratings from*

200 people regarding a drawing and concept statement used within an online survey at the early stage of packaging development.

Despite the limitations and challenges of using focus groups, there have been several published studies (Masson et al., 2016; Kobayashi and De Toledo Benassi, 2015; Metcalf et al., 2012; Pascall et al., 2009; Raz et al., 2008; Henson et al., 2006; Silayoi and Speece, 2004; Hill and Tilley, 2002) that have demonstrated the successful application of focus groups to explore and capture insights from consumers regarding packaging design and technologies. For example, Silyayoi and Speece (2004) utilized two focus groups to understand Thai consumers' shopping behavior toward food products and how the components of packaging affected buying decisions. The groups identified the packaging elements of graphics, color, shape, size, and product information as the most important and having the greatest impact on their purchasing decisions. The way in which the visual and informational packaging elements affect product selection and purchase will be strongly influenced by product involvement levels and time pressure at point of sale. Research by Silyayoi and Speece (2004) suggests that visual elements, graphics, size, and shape positively influence choice more in low-involvement purchase decisions, whereas informational elements tend to have a stronger role in higher-involvement purchase decisions.

Alternatively, Kobayashi and De Toledo Benassi (2015) utilized focus groups to evaluate six instant coffee packages which were selected based on diversity of color, shape, material, label information, price, and brand. The focus groups were encouraged to discuss and evaluate the visual characteristics to determine the attributes of most interest. Based on the qualitative results from the focus groups, some packaging attributes were selected for the development of the coffee packages images for the subsequent conjoint analysis study to determine the impact of packaging characteristics on consumers' purchase intention.

In contrast, the research of Pascall et al. (2009) demonstrates how focus groups with an educational component can be used to learn about consumer perceptions of new packaging technologies. Their study explored the use of focus groups to guide the development of a new tamper-evident packaging technology at its early conceptual stage so that appropriate changes to the device could be made and an understanding of consumer's experiences with tamper-evident devices could be obtained. Twelve focus groups were convened at six separate locations in the United States of America. Participants discussed their understanding and perceptions of tamper-evident devices, as well as being shown 16 different types of devices to encourage sharing of concerns or ideas about these devices. Pascall et al. (2009) concluded that the focus group method allowed a more thorough investigation of the participants' understanding and perceptions about specific products, as well as prior experiences with tamper-evident food packaging, that could not have been achieved via a survey or other similar quantitative method.

From the published studies reviewed, there appears to be a consensus that small-scale, qualitative methods and techniques, such as focus groups, word association, and sentence completion and/or construction, are useful and can contribute unique insights to optimize packaging design, especially at the early stages of the development process.

7.4   Consumers' Interaction With Packaging In-home and In-store: Observation

There is no single formula for uncovering opportunities and ideas for packaging innovation. However, there is a research technique that is often overlooked by researchers but utilized by packaging design practitioners, and that is the power of observation, both in-store and in-home. In-home observation is critical for packaging designers to use because often consumers within focus groups find it difficult to consistently visualize or verbalize the "next big packaging idea" for a product. However, their interactions with a product and its packaging within the home environment often demonstrate their feelings, for example, embarrassment or pleasure in having the product on view in a room, or how they accommodate problematic design elements in the way that they store or use a product. Because these behaviors develop over time and often become unconscious responses, consumers may no longer be aware of them, and therefore minor inconveniences such as difficulty of holding or pouring from a package are accepted without a second thought, and therefore often not raised as issues when asked directly, such as in one-to-one interviews or focus group situations. In essence, observational research is essential to understanding product and packaging usage in context, in particular for problems with current packaging structures.

This realization has given rise to an increase in ethnographic research, whereby practitioners learn by observing how consumers interact with packages, typically in a home environment. Usually, a small research team will go into a consumer's home and over time observe how individuals or families typically use, store, and interact with current or newly created packaging structures. Because the purpose of the ethnographic research is to spend time with participants in their own home environment, a deep understanding of their lifestyle or cultures is acquired and forms a basis for better understanding their needs and problems (Creusen et al., 2013). During the observational process, informal interviewing will be used to explore observed situations further to gain greater insight and complement the documentation process.

The most informative insights are often derived from documenting (photographs and/or videos) and understanding the full packaging life cycle (from purchase through transport, usage, storage, and ultimately disposal). In-home storage is particularly interesting because of the direct relationship between where and how long a product is stored (especially food and drink) with the context and speed with which it is consumed. Young (2004) cites the example of how in-home ethnographic research led to the development of "fridge packs" for beverages. The research demonstrated that carbonated soft drinks which "lived" in the fridge were consumed far more quickly than those kept in the garage or cupboard. In this instance, observational ethnographic research delivered important, unique information with convincing and presentable outcomes.

Store-based observation is critical to understand why so many packages are either opened or squeezed and then discarded. There is anecdotal evidence to suggest that consumers typically open or squeeze packaging to get a better "feel" for a product's dimensions, appearance, smell, and/or freshness, especially when there is no or limited product visibility through the packaging. In these instances, consumers are often trying to reassure themselves of product quality, particularly if the product is a high-involvement, infrequent purchase or is a gift to a friend or family member. These in-store assessments can be undertaken through accompanied shops or through in-store intercepts. Essentially, an accompanied shop involves an interviewer walking around a store with a shopper. The shopper is required to think out loud, thus verbalizing all their thoughts regarding various shopping activities. The interviewer will observe and probe to understand how and why they are choosing particular items and/ or responding to specific stores cues, for example, special offers or packaging. The information elicited from accompanied shops enables a holistic, naturalistic understanding of individual and/or group shopping behavior and how they interact with the shopping environment (Lowrey et al., 2005).

In addition to understanding how consumers interact with a product's packaging on shelf, there is a need to assess how packaging performs in terms of standout and engagement of consumers at point of product choice. Eye tracking (via a fixed or mobile system) can be utilized to measure consumers' response to a standalone package structure and graphics and/or in the context of shelf point of sale. Eye-tracking software indirectly measures consumers' attention and engagement with packaging through the tracking, recording, and analysis of participants' eye movements when looking at visual stimuli. The eye movements provide an objective indicator of where a participant's overt attention is focused. Piqueras-Fiszman et al. (2013) utilized the techniques of word association and eye tracking to collect attentional information and freely elicited associations from consumers in response to changing some specific features of jam jars, for example, jar shape (rounded versus square) and texture (smooth versus ridged). The eye-tracking aspect of the study revealed that certain elements of the packaging can be used to drive consumers' attention to a particular feature; for example, the ridged surface of the jars spread a participants' gaze to the border areas of the packaging, whereas a rounded jar seemed to direct participants' attention to the flavor label.

7.5 An Evaluation of the Relative Importance of Different Packaging Components and Extrinsic Cues: Conjoint Analysis and MaxDiff

7.5.1 Conjoint Analysis

Conjoint analysis is used to assess the relative importance of different packaging elements that enhance consumers' perception of the product. It is based on the principle of measuring consumers' trade-offs between discrete attributes to determine which features a packaging design should have. Rather than directly asking respondents what are the most important attributes, the method presents a series of options from which a respondent can only choose one each time, encouraging them to act instinctively on an intuitive basis. Consequently, by aggregating the utility scores of each attribute, only those attributes that are very strong will be identified.

Conjoint analysis has been widely used to evaluate consumer preferences of food product attributes (Koutsimanis et al., 2012; Ares and Adriana, 2007, and Jaeger, 2000) as well as packaging attributes (Loebnitz et al., 2015; Kobayashi and De Toledo Benassi, 2015; Georgios et al., 2012; Mueller Loose et al., 2013; Ares and Deliza, 2010b; Silayoi and Speece, 2005, and Raz et al., 2008). Saporta (1996) (cited in Raz et al. (2008)) succinctly characterizes the objectives of conjoint analysis:

- To measure the importance of each attribute for the definition of a packaging design;
- To measure the respective utilities of each level pertaining to an attribute. A level of an attribute can be: intensity, variation, or price point. For example, if "Fat Content" was a labeling attribute, it could have 30% fat content, 50% fat content, and 80% fat content as the levels;
- To describe and quantify prospective consumers for a packaging design. Conjoint analysis enables recreated or new attribute combinations to be submitted to a market simulator function to identify potential consumer choice and preference share. Moskowitz et al. (2009) demonstrate the ability of conjoint analysis to estimate the total utility value of each recreated concept by summing the utilities of the components;
- To determine the characteristics of an ideal packaging design for a group of consumers.

Conjoint analysis is best applied to situations where there are a number of attributes to be evaluated and each attribute has a number of discrete levels. The levels can be presented as text and/or images to the respondents, and they are typically known as elements. To obtain valid data, it is generally accepted that the way the conjoint task is presented, that is, the combination of text and/or images must closely resemble the way consumers make marketplace choices. The use of images can enhance task realism and external validity in product categories where consumer choice is highly influenced by visual inspection of the product and packaging prior to purchase. For example, Kobayashi and De Toledo Benassi (2015) conducted a conjoint analysis study to evaluate the purchase intention of different coffee packaging designs: glass jars and refill packs. For refill packs, they found that consumers were more willing to buy the coffee if there was a photograph of the coffee, coupled with additional information about the product. With regard to the glass jars, consumers' willingness to buy was dependent on the shape of the jar and the price of the coffee. Neither packaging types were influenced by brand, suggesting that packaging attributes and price are of greater importance to consumers. When evaluating packaging components and overall design, brand, and logos are not applied to minimize respondent bias and distraction from the task.

Typically a conjoint analysis is delivered online, and therefore the scale of the study needs to be taken into consideration in terms of completion time. This has an impact on the number of attributes and levels selected for the study. In studies completed by Georgios et al. (2012); Ares and Deliza, (2010b); Silayoi and Speece (2005); Chen and Burgess (2010); and Thomas (2015), the number of attributes and levels were restricted to 7 and 1, 2 and 3, 5 and 2, 8 and 4, and 5 and 3, respectively. Following the application of conjoint analysis to assess the sophisticated packaging for wine, Raz et al. (2008) acknowledge the difficulty in determining a suitable cut-off point when selecting appropriate attributes and the levels per attribute to be assessed, as the potential combinations could be very important and essential to the reliability of the results. Silayoi and Speece (2005) comment that there is a general consensus toward balancing the number of attributes required to realistically represent the package against the need to simplify the representation so that it does not overly complicate the task for the respondents.

Overall the published studies (Kobayashi and De Toledo Benassi, 2015; Loebnitz et al., 2015; Georgios et al., 2012; Ares and Deliza (2010b); Silayoi and Speece, 2005; Chen and Burgess, 2010; Thomas, 2015) would suggest that conjoint analysis is an effective way to examine how consumers evaluate the relative importance of different packaging features, such as price, shape, color, labels, and graphics, and how their decision-making impacts on their buying intentions or preferences. Manufacturers of packaged food and drink products can use it to help create effective packaging designs and strategies. Utilizing the packaging attributes/features perceived as important by consumers overall and then by market segment can provide very useful information to marketers who want to maximize the package's impact in selling the product.

7.5.2 MaxDiff

MaxDiff, also known as Best–Worst scaling, is a common marketing approach for modeling consumers' expectations by obtaining importance (or preference) scores for multiple items, which are typically product brands, image statements, product or packaging features, advertising or on-pack claims, full concepts, and benefit articulations. This measurement and scaling technique, invented by Jordan Louviere and his colleagues in 1987, was developed to alleviate respondents' fatigue by reducing the number of items typically shown during traditional, well-established paired comparison exercises (method of paired comparisons dating back to the early 1900s) for eliciting trade-offs among items (Sawtooth, 2013).

MaxDiff finds its origins/roots in conjoint analysis, which permits intra-attribute comparisons of levels but does not permit inter-attribute comparisons due to the scaling of the attributes being unique to each attribute (Sawtooth, 2003). MaxDiff, in contrast, permits both intra- and inter-item comparisons of levels by measuring attribute level utilities (or importance) on a common, unidimensional interval-level scale of importance based on nominal choice data. Most of the prior applications of MaxDiff have been within conjoint analysis (Best–Worst Conjoint Analysis), in which respondents are presented with a full product or packaging profile. MaxDiff shares much in common with conjoint analysis but is easier to use for both the researchers and the respondents and has a wider variety of applications: it may be used to prioritize packaging features (eg, packaging material versus pack size) that are most important to potential customers, identify key messages to be included on the packaging (eg, fat free versus sugar free), or identify the packaging prototypes with the greatest potential for success on the market (Lagerkvist, 2013).

Creating a MaxDiff task typically starts with the selection of relevant attributes/items to be assessed depending on the research objectives (12+), then an appropriate number of sets/subsets of items to be assessed by each respondent (10+) and a relevant number of items to be assessed with each set (typically four or five). Item allocation within and between sets is based on an incomplete block design, taking into account the number of test respondents (typically 300+ when segmentation is envisaged).

A good design controls for both order (each item is seen in each position order across subsets of items an equal number of times) and context (each item is seen with every other item an equal number of times) effects.

Instead of indicating which items are important using CATA scales, directly scoring or rating each item for importance using a continuous/line or a category scale, or instead of ranking all items to be compared in terms of importance, respondents are shown successive sets/subsets of the assessed items and are asked to indicate the best (or most important) and worst (or least important) attributes/items within each set. Respondents are typically shown a dozen sets of items. Attributes within each set are combined so that across all sets each item is shown ideally an equal number of times, typically more than twice, to each respondent.

This approach has numerous advantages. Standard (eg, 5-point) rating scales for importance tend to show that most items are important. Researchers showed that MaxDiff results bring about greater discrimination between the tested items and between the test respondents than when, for instance, scoring or rating scales are used. They also showed that MaxDiff is a reasonably simple task to perform by respondents from various educational and cultural backgrounds. Since respondents make choices rather than scoring their level of preference using a numeric scale, there is also no opportunity for scale use bias, which is particularly valuable when conducting cross-cultural studies. In addition, more items can be tested than with other direct testing approaches (eg, ranking and paired comparison). MaxDiff typically leads to a rank and to a metric distance between the items that indicates their relative importance (or preference). As a consequence, the items can be prioritized and the key/critical ones identified.

However, due to the task being fairly repetitive, drawbacks can be respondents' boredom and loss of concentration with too many sets of items presented. It is also less suited than conjoint analysis to studying products/packaging concepts made up of complex features added together and to handling pricing data. Data analysis can also be more difficult to handle with non-appropriate data capture and/or statistical software as they involve conditional probability models. For each item incorporated into the design, the simplest statistical treatment that can be applied to the data consists of counting how many times each item was chosen "best" or "worst." However, that approach is only sufficient for a top-line view of the data and valid as long as each item is displayed an equal number of times in a balanced experiment. A more sophisticated but widely accepted data treatment approach is the Hierarchical Bayes (HB) statistical technique to estimate the so-called MaxDiff scores (rescaled to 100 so the scores of all items sum up to 100), for each respondent and at an overall level (all respondents). To inform marketing strategies, segmentation analysis (typically latent class rather than cluster analysis) is commonly simultaneously applied alongside HB to unfold and specifically target segments of respondents with similar needs or preferences (Sawtooth, 2013).

In the literature, MaxDiff usage has been essentially reported in the context of a product and/or its packaging's concept definition or improvement (Lagerkvist, 2013), that is, where it is essential for industry to understand what influences the choice/purchase of their product(s) and as such what could be a "must have" feature to stand out from the competition. In such a context, product tasting is not involved and studies are mostly conducted as online surveys, sometimes utilizing product pictures (typi-cally in its packaging or as it would be typically bought by consumers) to bring some contextual elements.

Chambault (one of the authors), in a recent research study still unpublished to date (2015), has also used MaxDiff to define the perceptual (ie, sensory) profile of a set of products belonging to the same product category (hard cheese) and compare the products on their resulting profiles.

Other recent applications of MaxDiff have been in the area of sensory brand-ing, that is, where brand singularity and brand/product consonance or congruence are of particular interest. With food products, these studies may involve tasting the product(s) and may therefore be conducted at a test venue. Thomson and Crocker (2014, 2015) have used Best–Worst Scaling—as an alternative to indirect measure-ments of consumer response such as the measurement of consumer facial expres-sions and brain activity—to elicit the conceptual associations (also called implicit associations) that consumers make with brands, products, and packaging. In their studies, respondents were typically presented with the object/products to be profiled and asked to choose the words the most and the least associated with the object/product from successive sets of four or five words. The resulting conceptual profile took the form of a series of words placed on a difference scale, according to their degree of association with the object/product. According to the authors, all objects, including brands and products, have perceptual (ie, sensory) characteristics and con-ceptual associations. Together they determine how an object seems to us and how it impacts on our feelings. Capturing and quantifying the conceptual associations that trigger the feelings that induce reward and subsequently motivate behavior (through conceptual profiling) provides a rich source of insight for guiding brand and product development.

7.6   Holistic Approaches to Explore the Similarities and Differences in Overall Packaging and Packaging Design Components: Projective Mapping and Related Techniques

Due to their time- and cost-effectiveness for characterizing a large set of products from an objective (with sensory assessors) or a subjective (with consumers) point of view, rapid techniques such as napping or projective mapping, sorted napping, and (free) sorting have gained popularity among both sensory and consumer research prac-titioners when it comes to assessing the product itself or its packaging (as a whole or its constituents). They are often referred to as "holistic" approaches, that is, sponta-neous and less rational methods of data collection (than scoring or rating products on a predefined list of attributes or statements), whereby assessors are free to use their own criteria to discriminate products on an overall basis, in blind conditions (products are unbranded), in informed conditions (products are in their packaging), or based solely on their packaging (whole packaging or packaging elements).

Napping (Pagès, 2003 cited in Liu et al., 2016) is a specific variant of projective mapping, originally proposed for applied sensory studies by Risvik et al. (1997, cited in Liu et al., 2016) to identify overall differences and similarities within a set of products. The products are simultaneously presented to the assessors, who are then required to project them onto a two-dimensional space (materialized by a large sheet of paper or a defined area on a computer screen) in a way that reflects the differences they perceive between products: products perceived to be similar are placed close to each other and products perceived to be different are placed apart from each other. Projective mapping is often followed by a descriptive task named ultra-flash profiling (UFP), whereby assessors comment on each of the products or groups of products they identified. UFP is typically performed to help interpret the perceptual average/ group product map resulting from the analysis of the projective mapping data. Many researchers have been describing projective mapping as a relatively cheap, rapid, and user-friendly method to use (Albert et al., 2011 and Veinand et al., 2011 cited in Liu et al., 2016). However, recently this method has also been criticized by other researchers for its low reliability (ie, low discrimination and repeatability) without a short training session either on the method itself or on the characteristics expected to be in the tested products (Liu et al., 2016).

A simpler alternative to projective mapping, free sorting, only requires the assessors to split the products into different groups (according to their dominating features) and then to describe the groups. Halfway between napping and free sorting, that is, with an emphasis on both relative distances between the products and product grouping, is sorted napping, which involves assessors positioning products on a two-dimensional surface according to their similarities and differences, then grouping similar products together and describing each of the groups formed.

Chambault (2013) did a comparison between napping, sorting, and sorted napping on the same product set and found them particularly useful to support benchmarking and R&D initiatives (eg, product formulation) or reduce the size of a large set of products belonging to the same category prior to more in-depth (sensory or consumer) analysis. These holistic approaches, however, remain exploratory in nature: they cannot be used to individually characterize each of the products or draw conclusions regarding significant differences between the products within the assessed set.

In the context of product evaluation, many applications of such methods have been found with sensory assessors or professionals (objective product evaluation), particularly within the beverage (ie, wine) industry. In recent years, more applications have been found with consumers (subjective evaluation) to assess the product itself (typically unbranded, ie, in a blind condition), its packaging (expected condition) or both the product and its packaging (in informed conditions). Hereafter are some of the most recent applications found in the literature, where products were assessed based on their whole packaging or some of their components (eg, key packaging information shown on experimental cards and packaging labels, logos, and claims, focusing on, eg, contents, size, shape, color, and font), often to show that expectations built from the packaging impact on the perception of the product itself (eg, sensory characteristics, liking, perceive quality, and healthiness).

Boraud et al. (2011) used what they called "personalised holistic approaches" to capture the consumer perception of 16 wines based on their labels: consumers could choose to perform either a free sorting task or a sorted napping task. Using hierarchical multivariate factor analysis (HMFA), the authors demonstrated that both methods led to the same criteria of label separation: design, followed by appellation, vintage, and price. Chen and Chambault (2013) used projective mapping with free elicited descriptions to characterize 14 wines based on their whole packaging (ie, their whole bottles as opposed to their labels only). The authors confirmed that projective mapping could be an effective tool to quickly assess not only food but also non-food items (product packaging) and to identify the key packaging attributes that are relevant to consumers of different demographics and behaviors. Perceived quality of wine, liking of the product or its packaging, and perceived overall design differentiation were found to be the three main criteria used by all groups of consumers investigated to discriminate between wine packaging.

Carrillo et al. (2012a) used projective mapping to assess the combined effect of product non-sensory (packaging information) and sensory characteristics on the perception (healthiness and acceptability) of 10 energy-enriched or -reduced biscuits in three different scenarios: in blind (products unbranded), informed (products in their packaging), and expected (product packaging) conditions. The results showed that acceptance was generally higher when the biscuits were assessed in the blind condition, and that there was a trend for the biscuits to be perceived as healthier when assessed in the expected condition. Carrillo et al. (2012b) also conducted projective mapping in two packaging (whole packaging versus experimental cards with the packaging nutrition- and claim-related information) and two product tasting (with and without nutrition- and claim-related information) conditions to ascertain how consumers perceive and use the nutrition information panel, as well as the nutrition- and health-related claims among other packaging cues to discriminate between 23 energy-enriched or -reduced biscuits. Among all packaging cues, consumers were found to be greatly influenced by the claims (the nutritional ones in particular) on the front of the packaging, and the product information had a negative influence on the perception of their hedonic sensory characteristics.

When projective mapping only involves assessing the product packaging and not its influence on the product (in which case the product does not need to be assessed), advances in data capture technology allow this technique to be conducted online, using packaging images, where a larger pool of consumers can be used than when conducting the test at a venue. Chambault (2015) used projective mapping as part of an online survey on cereal bars to assess consumers' behavior toward sustainable foods based on their overall perception, familiarity, and understanding of 25 food logos. The research showed how consumers interpret food logos, how they distinguish between sustainable and other logos, how consumers' interpretation of food logos, overall concern, and understanding about the ecological issues and attitudes to health influence their food choices, and most importantly, that logos consumers can directly relate to product quality and/or satisfaction could have a greater impact on their behavior.

7.7 Conclusions

This chapter has presented a series of explicit qualitative and quantitative methods to investigate consumers' perceptions, attitudes, feelings, and experiences of packaging concepts, designs, and technologies. Although for the purposes of this chapter these methods have been presented singularly, many of the studies reviewed have employed a multimethod approach. For example, Raz et al.'s study (2008) utilized an integrated approach that started with qualitative focus group exploration, followed by quantitative hedonic assessment and conjoint analysis; Piqueras-Fiszman et al. (2013) combined eye tracking with word association to assess novel packaging solutions; Ares and Deliza (2010a) utilized the techniques of free listing and word association to identify important package features of milk desserts; and Ampuero and Vila (2006) applied interviewing and napping to elicit and quantify consumers' perceptions of product packaging. These and other studies have demonstrated the synergistic benefits derived from triangulating methods when studying phenomena of packaging from the consumer perspective.

The studies presented in this chapter have also demonstrated that consumer research is fundamental to the development of successful packaging in terms of its structural, graphical, functional, and symbolic properties. Consumer research provides a greater depth of understanding; it elicits what consumers truly think and feel about a product's packaging, as well as how it is used and interacted with. To communicate the product experience effectively and optimize the potential of packaging, brand owners and marketers must understand consumer responses to their packages, and integrate the perceptual evaluations of consumers into the packaging design process.

References

Albert, A., Varela, P., Salvador, A., Hough, G., Fiszman, S., 2011. Overcoming the issues in the sensory description of hot served food with a complex texture. Application of QDA®, flash profiling and projective mapping using panels with different degrees of training. Food Quality and Preference 22, 463–473.

Ampuero, O., Vila, N., 2006. Consumer perceptions of product packaging. Journal of Consumer Marketing 23 (2), 100–112.

Arboleda Ana, M., Arce-Lopera, C., 2015. Quantitative analysis of product categorization in soft drinks using bottle silhouettes. Food Quality and Preference 45, 1–10.

Ares, G., Deliza, R., 2010a. Identifying important package features of milk desserts using free listing and word association. Food Quality and Preference 21, 621–628.

Ares, G., Deliza, R., 2010b. Studying the influence of package shape and colour on consumer expectations of milk desserts using word association and conjoint analysis. Food Quality and Preference 21, 930–937.

Ares, G., Adriana, G., 2007. Influence of gender, age and motives underlying food choice on perceived healthiness and willingness to try functional foods. Appetite 49, 148–158.

Becker, L., et al., 2011. Tough package, strong taste: the influence of packaging design on taste impressions and product evaluations. Food Quality and Preference 22 (1), 17–23.

Boraud, B., Cadiou, H., Druon, A., Le, S., Pages, J., 2011. Personalized Holistic Approaches Handled by Hierarchical Multiple Factor Analysis (HMFA). Example in Which Free Choice Concerning the Kind of Questionnaire Is Given to Consumers. Poster available at: http://carme2011.agrocampus-ouest.fr/poster/PersonalizedHolisticApproaches.pdf.

Burgess, P., 2015. Consumer engagement: integrating packaging with the product experience. In: 11th Pangborn Poster and Oral Presentation.

Carr, et al., 2001. A case study in relating sensory descriptive data to product concept fit and consumer vocabulary. Food Quality and Preference 12, 407–412.

Carrillo, E., et al., 2012a. Effects of food package information and sensory characteristics on the perception of healthiness and the acceptability of enriched biscuits. Food Research International 48, 209–216.

Carrillo, E., Varela, P., Fiszman, S., 2012b. Packaging information as a modulator of consumers' perception of enriched and reduced-calorie biscuits in tasting and non-tasting tests. Food Quality and Preference 25 (2), 105–115.

Chambault, M., 2015. Encouraging Positive Attitude/Behaviour Change towards Sustainable Foods: Online, Direct and Indirect Measurements of Consumers' Responses to Food Logos. Campden BRI R&D. report 393.

Chambault, M., 2013. Sorting, Sorted Napping and Napping: How They Can Complement Conventional Sensory Profiling Methods to Support R&D Initiatives. Campden BRI R&D. report 342.

Chen, X., 2014. Assessing Dissonance between Packaging and Product Experience: A White Wine Study. Campden BRI R&D. report 379.

Chen, X., Chambault, M., 2013. Assessing Wine Packaging Using the Napping Technique. Part of the Research Project An Integrated Approach to Sensory Evaluation of Product and Packaging. Campden BRI R&D. report 355.

Chen, X., Burgess, P., 2010. The Relative Importance of Environmental Concerns and Ethical Practices when Consumers Make Food and Drink Purchasing Decisions. Campden BRI R&D. report 293.

Creusen, M., Hultink, E.J., Eling, K., 2013. Choice of consumer research methods in the front end of new product development. International Journal of Market Research 55 (1), 81–104.

Delgado, C., et al., 2013. Evaluating bottles and labels versus tasting the oils blind: effects of packaging and labelling on consumer preferences, purchase intentions and expectations for extra virgin olive oil. Food Research International 54 (2), 2112–2121.

Donoghue, S., 2000. Projective techniques in consumer research. Journal of Family Ecology and Consumer Sciences 28, 47–53.

Fenko, A., Kersten, L., Bialkova, S., 2016. Overcoming consumer scepticism toward food labels: the role of multisensory experience. Food Quality and Preference 48, 81–92.

Georgios, K., Getter, K., Behe, B., Harte, J., Almenar, E., 2012. Influences of packaging attributes on consumer purchase decisions for fresh produce. Appetite 59, 270–280.

Gordon, W., Langmaid, R., 1988. Qualitative Market Research. Gower, Aldershot.

Guerrero, L., Laret, A., Verbeke, W., Enderli, G., Zakowska-Biemans, S., Vanhonacker, F., Issanchou, S., Sajdakowska, M., Grali, B.S., Scalvedi, L., Contel, M., Hersleth, M., 2010. Perception of traditional food products in six European regions using free word association. Food Quality and Preference 21, 225–233.

Gutjar, S., 2014. Food choice: the battle between package, taste and consumption situation. Appetite 80, 109–113.

Henson, B., Barnes, C., Livesey, R., Childs, T., Ewart, K., September 2006. Affective consumer requirements: a case study of moisturiser packaging. Concurrent Engineering 14 (3), 187–196.

Hill, H., Tilley, J., 2002. Packaging of children's breakfast cereal: manufacturers versus children. British Food Journal 104 (9), 766–777.

Jaeger, S.R., 2000. Uncovering cultural differences in choice behaviour between Samoan and New Zealand consumers: a case study with apples. Food Quality and Preference 11, 405–417.

Kinnear, T.C., Taylor, J.R., 1991. Marketing research: an applied approach. McGraw-Hill, London.

Kobayashi, M.L., De Toledo Benassi, M., 2015. Impact of packaging characteristics on consumer purchase intention: instant coffee in refill packs and glass jars. Journal of Sensory Studies 30, 169–180.

Koutsimanis, G., Getter, K., Behe, B., Harte, J., Almenar, E., 2012. Influences of packaging attributes on consumer purchase decisions for fresh produce. Appetite 59, 270–280.

Labbe, D., Pineau, N., Martin, N., 2013. Food expected naturalness: impact of visual, tactile and auditory packaging material properties and role of perceptual interactions. Food Quality and Preference 27 (2), 170–178.

Lagerkvist, C.J., 2013. Consumer preferences for food labelling attributes: comparing direct ranking and best-worst scaling for measurement of attribute importance, preference intensity and attribute dominance. Food Quality and Preference 29, 77–88.

Liao, L.X., et al., 2015. Responses towards food packaging: a joint application of self-report and physiological measures of emotion. Food Quality and Preference 42, 48–55.

Liu, J., et al., 2016. Performance of flash profile and napping with and without training for describing small sensory differences in a model wine. Food Quality and Preference 48, 41–49.

Loebnitz, N., Mueller Loose, S., Grunert, K., 2015. Impacts of situational factors on process attribute uses for food purchases. Food Quality and Preference 44, 84–91.

Lowrey, T.M., Otnes, C.C., McGrath, M.A., 2005. Shopping with consumers: reflections and innovations. Qualitative Market Research 8 (2), 176–188.

Masson, M., Delarue, J., Bouillot, S., Sieffermann, J.M., Blumenthal, D., 2016. Beyond sensory characteristics, how can we identify subjective dimensions? A comparison of six qualitative methods relative to a case study on coffee cups. Food Quality and Preference 47, 156–165.

Metcalf, L., Hess, J.S., Danes, J.E., Singh, J., 2012. A mixed-methods approach for designing market-driven packaging. Qualitative Market Research 15 (3), 268–289.

Moskowitz, H., Reiner, M., Lawlor, J.B., Deliza, R., 2009. Packaging Research in Food Product Design and Development. Wiley-Blackwell, Hoboken.

Mueller, S., Szolnoki, G., 2010. The relative influence of packaging, labelling, branding and sensory attributes on liking and purchase intent: consumers differ in their responsiveness. Food Quality and Preference 21 (7), 774–783.

Mueller Loose, S., Peschel, A., Grebitus, C., 2013. Quantifying effects of convenience and product packaging on consumer preferences and market share of seafood products: the case of oysters. Food Quality and Preference 28, 492–504.

Ngo, M.K., Piqueras-Fiszman, B., Spence, C., 2012. On the colour and shape of still and sparkling water: Insights from online and laboratory-based testing. Food Quality and Preference 24, 260–268.

Pascall, M.A., Lee, K., Fraser, A., Halim, L., 2009. Using focus groups to study consumer understanding and experiences with tamper-evident packaging devices. Journal of Food Science Education 8, 53–59.

Piqueras-Fiszman, B., Alcaide, J., Roura, E., Spence, C., 2012. Is it the plate or is it the food? Assessing the influence of the color (black or white) and shape of the plate on the perception of the food placed on it. Food Quality and Preference 24, 205–208.

Piqueras-Fiszman, B., Velasco, C., Salgado-Montejo, A., Spence, C., 2013. Using combined eye tracking and word association in order to assess novel packaging solutions: a case study involving jam jars. Food Quality and Preference 28, 328–338.

Raz, C., Piper, D., Haller, R., Nicod, H., Dusart, N., Giboreau, A., 2008. From sensory marketing to sensory design: how to drive formulation using consumers' input? Food Quality and Preference 19, 719–726.

Risvik, E., McEwan, J.A., Rødbotten, M., 1997. Evaluation of sensory profiling and projective mapping data. Food Quality and Preference 8, 63–71.

Roininen, K., Arvola, A., Lahteenmaki, L., 2006. Exploring consumers' perceptions of local food with two different techniques: laddering and word association. Food Quality and Preference 17, 20–30.

Saporta, G., 1996. Methode et applications de l'analyse conjointe. In: Sino-French Workshop on Advanced Data Analysis Methods in Industry and Management. Beijing, cited in Raz, C., Piper, D., Haller, R., Nicod, H., Dusart, N., Giboreau, A., 2008. From sensory marketing to sensory design: How to drive formulation using consumers' input? Food Quality and Preference 19, 719–726.

Sawtooth Software (Steven H. Cohen, SHC & Associates), 2003. Research Paper Series. Maximum Difference Scaling: Improved Measures of Importance and Preference for Segmentation.

Sawtooth Software, 2013. Technical Paper Series. The MaxDiff System. Technical Paper, version 8.

Schouteten, et al., 2015. Linking emotions and sensory attributes of traditional food products. An application of the CATA approach in a study on children. In: 11th Pangborn Conference Poster.

Silayoi, P., Speece, M., 2005. The importance of packaging attributes: a conjoint analysis approach. European Journal of Marketing 41 (11/12), 1495–1517.

Silayoi, P., Speece, M., 2004. Packaging and purchase decision: an exploratory study on the impact of involvement level and time pressure. British Food Journal 106 (8), 607–628.

Thomas, S., 2015. Information on Fat and Sugar Content Together with Price Is Driving Consumers' Decision Making in the Context of Strawberry Yoghurt. Campden BRI R&D. report 381.

Thomson, D.M.H., Crocker, C., 2015. Application of conceptual profiling in brand, packaging and product development. Food Quality and Preference 40 (Part B), 343–353.

Thomson, D.M.H., Crocker, C., 2014. Anchored scaling in best worst experiments: a process for facilitating comparison of conceptual profiles. Food Quality and Preference 33, 37–53.

Tu, Y., et al., 2015. Touching tastes: the haptic perception transfer of liquid food packaging materials. Food Quality and Preference 39, 124–130.

Tuckman, B.W., 1965. Developmental sequence in small groups. Psychological Bulletin 63 (6), 384–399.

Vadiveloo, M., et al., 2013. The interplay of health claims and taste importance on food consumption and self-reported satiety. Appetite 71, 349–356.

Vasiljevic, M., et al., 2015. Making food labels social: the impact of colour of nutritional labels and injunctive norms on perceptions and choice of snack foods. Appetite 91, 56–63.

Veinand, B., Godefroy, C., Adam, C., Delarue, J., 2011. Highlight of important product characterists for consumers. Comparison of three sensory descriptive methods performed by consumers. Food Quality and Preference 22, 474–485.

Westerman, S.J., et al., 2013. The design of consumer packaging: effects of manipulations of shape, orientation, and alignment of graphical forms on consumers' assessments. Food Quality and Preference 27, 8–17.

Will, V., Eadie, D., MacAskill, S., 1996. Projective and enabling techniques explored. Marketing Intelligence and Planning 14 (6), 38–43.

Young, S., 2004. Breaking down the barriers to packaging innovation. Design Management Review 15, 68–73, Winter.

Consumers' Mindset: Expectations, Experience, and Satisfaction

P. Burgess
Campden BRI, Chipping Campden, United Kingdom

8.1 Consumer Choices

Packaging has to work hard at a number of different levels, not least in importantly shaping and informing consumer expectations regarding the product itself but also in driving product selection in highly populated product categories.

The choice explosion that characterizes many grocery product categories was discussed in a paper by Schwartz and Ward (2004), who highlighted that a typical US supermarket will often now carry more than 30,000 items. For example, in one retailer they examined, they noted that there were 85 different varieties and brands of crackers, 286 varieties of cookies, 75 iced teas, 40 options for toothpaste, 360 types of shampoo, conditioner, gel, and mousse, and 120 different pasta sauces.

Although significant resources are allocated in ensuring that product assortments within a retail environment are profitably optimized, there is evidence that this "choice" architecture is perhaps not as fully reflective of consumer needs as is often thought. (Indeed at the time of writing, many grocery retailers in the UK are critically reappraising their category assortments in response to market pressures and changing consumer shopping patterns.)

As Schwartz and Ward go on to comment, the fact that some choice is good does not necessarily mean that more choice is better; in other words, there is a cost to having an overabundance of choice, particularly in relation to the "cognitive load" that this choice can bring to the purchase decision.

So how do consumers navigate their way through this sea of choices?

Over the past decade, there have been significant developments in insights gained from social psychology in which the central theme is the relationship between conscious and unconscious elements in behavior and decision-making (see Spinelli and Niedziela, chapter: Emotion Measurements and Application to Product and Packaging Development).

Many of these insights underpin the emergence of behavioral economics, which links economic theory to social and environmental psychology. Work published by academics, particularly Kahneman (2003), has made behavioral economics highly visible and relevant to commercial organizations and those whose aim is to change society's behavior.

Behavioral economics questions the premise that the choices consumers make are the result of deliberative, linear, and controlled processes, and instead asserts that consumers are strongly influenced by context; decisions are guided by conditions of the moment—by social and environmental influences as well as by cognitive shortcuts, emotions, and habits.

Integrating the Packaging and Product Experience in Food and Beverages. http://dx.doi.org/10.1016/B978-0-08-100356-5.00008-5

Kahneman (2003) notably discusses the developments in research on intuitive thinking and decision-making. In this paper, the dual system view is discussed:

Intuition (System 1 thinking) is characterized by operations that are "fast, parallel, automatic, effortless, associative, implicit, and often emotionally charged; they are also governed by habit and are therefore difficult to control or modify."

In contrast operations under System 2 thinking are "slower, serial, effortful, more likely to be consciously monitored and deliberately controlled; they are also relatively flexible and potentially rule governed."

Importantly, according to Kahneman, intuitive processes deal with the large majority of problems and choice decisions on a day-to-day basis, and although it is one of the functions of System 2 to monitor the quality of both types of mental operations, it often neglects this task since it is susceptible to interference by other effortful tasks.

Some of these themes are discussed by Köster (2009). For example, Köster highlights the role of incidental (nonintentional) and unconscious learning in food choice and preference. Köster notes that through such incidental learning, a large amount of information is gathered and stored by individuals so that they are not consciously aware of. This information often only becomes evident when the expectations that are built upon it are not met with future encounters with similar experiences or situations.

Köster notes that even when respondents explicitly and consciously try to remember the specific taste of a previously eaten food or even to recognize it among other variants of that same food, they usually do not succeed in finding the right one at a better than chance level, but respondents are quite good at rejecting variations of it as not meeting their "sensory" expectations.

One can immediately see how, in a retail environment, the visual aspects of packaging design such as color, form, or image mold (see Spence, chapter: Multisensory Packaging Design: Color, Shape, Texture, Sound, and Smell) will play a critical part in enabling the application of consumers' intuitive decision-making processes as well as creating/confirming expectations regarding the product experience.

As noted by Kahneman, System 1 thinking reinforces consumers' preference for things to remain the same, a tendency not to change behavior unless the incentive to do so is very strong—a status quo bias first described by Samuelson and Zeckhauser (1988).

A status quo bias leads to consumers ignoring choices available to them in favor of behavior established through routine or habit. Habits are automatic and appear as instinctive and usually develop when actions are repeatedly paired with an event or context. There is therefore a tendency for consumers to "go with the flow," along with an inclination to be influenced by social norms and opt for "default" behavior.

What Kahneman and Tversky (1979) demonstrated through their Prospect theory was that loss aversion is a powerful influence underpinning this status quo bias. Kahneman calculated that the prospect of a subjective loss is up to twice as powerful in preventing change than that of a subjective gain is in promoting it. In other words, in any decision, not doing badly is up to twice as significant as doing well.

Status quo bias and loss aversion work strongly together to steer consumer choices toward an established familiar product repertoire that enables consumers' more typical automatic decision-making processes to be effectively deployed.

The influences of status quo bias and loss aversion may explain some of the behavioral data that was reported in an article published in the April 2011 issue of the *Harvard Business Review*. The article noted that American families, on average, repeatedly bought the same 150 grocery items, which constituted as much as 85% of their household needs. (The same article went on to note that less than 3% of new consumer packaged goods exceeded first-year sales of $50 million—considered the benchmark of a highly successful product launch.)

Against this backdrop, we can see that, given the intransigence of consumer shopping habits, launching a new product is a risky business and, as summarized by Gordon (2011), "the body of evidence that has been built up under behavioral economics emphasises that human choice is not made deliberatively and consciously by weighing up and evaluating all the possible variables and permutations. Consumers make choices comparatively rather than absolutely, from what is available rather than taking a census, and in terms of how it makes them feel."

8.2 Expectations and Satisfaction

With a predisposition to the status quo, what triggers consumers' behavioral change?

Dannemiller and Jacobs (1992) looked at the problem of resistance to change and noted that three things are required, based on a simplified expression of the Gleicher change formula as follows:

$$D \times V \times F \text{ needs to be greater than } R$$

where D is dissatisfaction with the status quo, V is the vision of an alternative, and F is easy first steps. The three multiplied together need to be greater than R, resistance to change.

Although the approach was developed in the context of organizational change, the principles are readily applicable to consumer behavior in FMCG markets. With this behavioral change formula in mind, understanding consumer (dis)satisfaction of the consumption experience would therefore seem to be an important step in understanding consumer purchase behavior in terms of trial and/or repeat purchase of products.

So what is meant by (dis)satisfaction? The frameworks referenced in the satisfaction literature generally imply a conscious comparison between a cognitive state prior to an event and a subsequent cognitive state, usually realized after the event is experienced.

Giese and Cote (2000) characterized satisfaction as "a summary affective response of varying intensity, with a time-specific point of determination and limited duration, directed toward focal aspects of product acquisition and/or consumption."

As an affective dimension, the construct of satisfaction could be regarded as similar to other affective and/or evaluative responses to food and consumer products, and as noted by Cardello (2007), perhaps the liking/disliking of food can be analyzed using the same models that have been applied to product satisfaction and may therefore offer a more comprehensive insight on product acceptability than liking measures alone.

The Expectancy-Disconfirmation model offers a promising theoretical model for the assessment of consumption satisfaction. This theory underpins the assimilation-contrast approach that has been used by food science researchers as a basis to account for expectations-based effects on the response of consumers to a variety of food and drink products (eg, Yeomans et al., 2008) (Fig. 8.1).

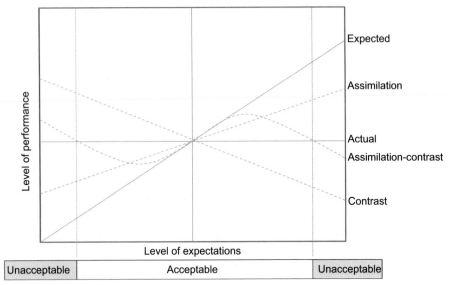

Figure 8.1 Representation of assimilation-contrast approach.
Adapted from Schifferstein et al. (1999)

The model implies that consumers select packaged food and beverage products with prepurchase expectations about the anticipated performance. The expectation level then becomes the standard against which the product is judged.

According to the model, consumption satisfaction for a packaged product depends on the degree of difference, or disconfirmation, between expectations and outcomes. Under the assimilation-contrast approach, if the degree of difference between what was expected and what was experienced is relatively small, the discrepancy will be minimized or assimilated by the consumer by changing his/her perception of the product to bring it more into line with his/her expectations.

(This effect is closely related to Festinger's (1957) theory of cognitive dissonance, defined as a psychologically uncomfortable state regarding potential future consumption outcomes. Because of the associated stress, various coping strategies, including the assimilation approach noted previously, are understood to be used by consumers to reduce dissonance.)

However, if the discrepancy between expectation and actual performance is so large that it falls into the zone of rejection, then a contrast effect comes into play, and the consumer exaggerates the perceived disparity between the product and his/her expectation of it.

An example of assimilation-contrast effect was reported in a study on white wines (Chen, 2014). In this study, four white wines were assessed by consumers at different time points under blind, packaging-evoked expectation, and informed conditions.

The first three columns in Table 8.1 show the mean scores with respect to consumers' overall liking of the products under blind, packaging, and informed conditions.

Table 8.1 **Assimilation-Contrast Results in a White Wine Study**

	Blind	Packaging	Informed	B–P	I–B	I–P
King's Favour	6.02	6.12	5.90	–0.10 NS	–0.12	–0.22
Nautilus	5.92	6.78	6.25	–0.85*** Disconfirmation	0.33 NS	–0.53
Vavasour	5.83	6.43	5.92	–0.60** Disconfirmation	0.09 NS	–0.50
Fairbourne	5.69	6.81	6.43	–1.12*** Disconfirmation	0.74** Assimilation	–0.37 NS, Complete

**p<0.01; ***p<0.001.
Chen, X.M., 2014. Assessing Dissonance Between Packaging and Product Experience: A White Wine Study. Campden BRI R&D Report No 379.

Under the blind condition, there was no significant difference between the four samples with respect to consumers' overall liking.

Under the packaging condition, Nautilus and Fairbourne were significantly ($p < 0.001$) more liked than King's Favour. Similarly, mean scores from the informed condition suggested that Nautilus and Fairbourne were liked more than King's Favour.

To establish whether the packaging had an impact on consumers' liking of the products, t-tests were performed on differences between blind and packaging hedonic ratings (B–P) for all samples. When the difference was significantly different from zero ($p < 0.05$), a disconfirmation occurs, which is an indication of the influence of packaging on consumers' liking of the product samples.

From Table 8.2, it can be seen that disconfirmation occurred in respect of the Nautilus, Vavasour, and Fairbourne samples. Thus, for King's Favour, the liking under the blind condition aligned with the expected liking under the packaging condition. For those samples with an observed disconfirmation, further t-tests were carried out on the differences between the informed and blind hedonic ratings (I–B) to establish whether the packaging and blind taste disconfirmation was associated with assimilation or a contrast effect.

An assimilation effect is an indication of a positive packaging impact on consumers' liking and a contrast effect suggests a negative packaging impact, which may lead to consumer rejection of the product. In this case, as shown in Table 8.2, the informed liking score for Fairbourne was significantly higher than the blind liking score, suggesting that an assimilation effect occurred; thus, a positive packaging impact was observed for this product.

In the assimilation case for Fairbourne, a t-test was then carried out on the difference between packaging and informed (P-I) liking scores to determine if the assimilation

Table 8.2 **Proposed Items in Generalized Consumption Satisfaction Scale**

1. This is one of the best (xyz) products I could have bought *(overall performance evaluation and quality)*.
2. This product is exactly what I needed *(need fulfillment)*.
3. This (xyz) product has not worked out as well as I thought it would *(failed expectations)*.
4. I am satisfied with my decision to buy this (xyz) product *(satisfaction "anchor")*.
5. I have mixed feelings about choosing this (xyz) product again *(cognitive dissonance)*.
6. My choice to buy this (xyz) product was a wise one *(success attribution)*.
7. If I could do it again, I would choose a different product *(regret)*.
8. I have truly enjoyed this (xyz) product *(positive affect/emotion)*.
9. I feel bad about my decision to buy this (xyz) product *(failure attribution)*.
10. I am not happy that I bought this (xyz) product *(negative affect/emotion)*.
11. Using this (xyz) product has been a good experience *(purchase evaluation)*.
12. I am sure it was the right thing to choose this (xyz) product *(success attribution)*.

effect is complete or not. As shown in Table 8.2, the difference between packaging and informed liking scores was not significantly different from zero; thus, the assimilation effect can be regarded as complete.

Despite its widespread popularity, the Expectancy-Disconfirmation model is not free of shortcomings. The main criticisms of this approach focus on the variability of expectations in terms of completeness, strength, and stability (Halstead et al., 1994).

For example, the suggested sequence of the model presupposes that everyone has precise expectations prior to the consumption experience. The model may be less meaningful in situations where consumers do not know what to expect until they experience the product. So while experienced consumers may rely on past consumption to shape their expectations, inexperienced consumers may rely more on external sources of information such as on-pack messages or referent groups to inform their views on what the product is likely to deliver.

Researchers in the field of consumption satisfaction also consider the Expectancy-Disconfirmation model to be too narrow cast in scope, and comment on how the intervening roles of purchase apprehension (eg, is this the best product I could have chosen?), purchase attribution (eg, was my choice of product a wise decision?), affect (eg, how do I feel about my product choice?), and need fulfillment play in enhancing, suppressing, or otherwise modifying the initial postpurchase responses of observing performance and making performance comparisons to a standard.

8.2.1 Purchase Apprehension

Many would argue that, along with increased choice in terms of products and purchase channels, there is a pressure to "maximize" the consumption outcome, that is, to seek the very best option available in a wide range of choice domains. However, it may well be the case that, for certain individuals, adding more choices simply makes choice more difficult, as they feel the pressure to choose the best possible option from an overwhelming array of choices. Under a maximizing strategy, the more choices that are presented, the greater the potential is for the consumer to experience "regret" at having chosen suboptimally.

As noted by Gilovich and Medvec (1995) and others, regret can play a powerful mediating role in the link between product choice and satisfaction with the consumption outcome. Researchers have noted that regret can be realized in both a "post-decision" context, that is, the feeling that rejected alternatives were better than the product chosen or that there were better alternatives out there that were not explored, but also as "anticipated" regret, that is, that a better alternative to the chosen product may be available at a foreseeable future date. Anticipated regret may produce not only dissatisfaction but also may completely inhibit the purchase decision.

Other individuals may aim to minimize potential regret by adopting a satisficing strategy, that is to choose the first option that surpasses some absolute threshold of acceptability, rather than attempt to optimize and find the best possible choice. This reflects the comments made by Gordon noted earlier in the chapter in that consumers, rather than engaging in an exhaustive search for information regarding a particular choice, would simply end their search as soon as an option was found that exceeded some criterion.

8.2.2 Purchase Attribution

Attribution processing, or in other words, assigning blame or praise for the consumption outcome, is a common phenomenon in reaction to consumption experiences.

The thinking behind this construct is that, when the product experience does not match consumers' prior expectations or other standards, consumers engage in an attributional process to make sense of what has occurred (Bitner, 1990). More specifically, the approach assumes that consumers tend to look for causes for product successes or failures and usually attribute these successes or failures across three dimensions:

- a locus dimension, that is, did the outcome reflect something done by the individual or something someone else did,
- a stability dimension, that is, is it expected that the outcome will always happen this way or will it vary over time,
- a controllability dimension, that is, was the outcome intended or unintended.

In the social psychology literature, the most prominent causal agent for the onset of attribution is the disconfirmation of expectations. In short, all forms of unexpectedness are likely to lead to attributional processing and can be regarded as an attempt by the consumer to understand the reasons behind the unexpected event so that the likelihood of potential negative surprise can be avoided in the future.

Attributional processing can therefore play a mediating role in the link between satisfaction and the disconfirmation of expectations.

8.2.3 Affect

Additionally, the more severe the consumption outcome, whether it be positive or negative, the greater the likelihood of attributional processing and degree of emotion felt.

Psychologists have shown that emotional associations and evaluations can have strong influences on consumption satisfaction and purchase decisions. Emotions can unconsciously or consciously reduce the complexity of decision-making, influencing automatic associations that guide consumers toward a decision. This occurs by means of "somatic markers," associations between stimuli and affective states that are based on past experience.

Capturing the emotional response to the consumption experience has grown in interest among product researchers in recent years. Some of these developments include the deployment of implicit neurometric-based approaches including, for example, facial coding, while others utilize explicit measurements such as the EsSense profile questionnaire framework developed by King and Meiselman.

These initiatives underpin the recognition that an emotional dimension represents an important component of the consumption satisfaction judgment (see Spinelli and Niedziela, chapter: Emotion Measurements and Application to Product and Packaging Development).

8.2.4 Need Fulfillment

Researchers also note the role of need fulfillment in shaping consumption satisfaction, as highlighted by the Value-Percept theory. According to this theory, satisfaction is an emotional response that is triggered by a cognitive evaluation process in which consumer perceptions of the product are compared to values, needs, wants, or desires (Westbrook and Reilly, 1983). Similar to the Expectancy-Disconfirmation model, a growing disparity between perceptions and values indicates an increasing level of dissatisfaction.

Discussion in the literature regarding the relationship between product attributes, need fulfillment, and satisfaction typically identifies three attribute categories:

- Bivalent satisfiers: the upward and downward translatable attributes that can cause both satisfaction and dissatisfaction,
- Monovalent dissatisfiers: essential but unprocessed attributes only capable of causing dissatisfaction when absent/not delivered,
- Monovalent satisfiers: additional, desirable "extras" that are processed at a higher level in the need hierarchy.

The Kano model, often referenced in product development frameworks, is an example of this approach to the classification of attributes and their relationship to satisfaction outcomes.

Researchers comment that disconfirmation of expectations and need fulfillment can both relate to satisfaction in an independent and additive fashion. In other words, falling short of, meeting, or exceeding needs may give different predictions of satisfaction than the question of falling short of, meeting, or exceeding expectations.

Hence several studies investigating the ability of value and expectations in determining satisfaction have demonstrated that it might be better to integrate needs and expectations into a single framework (Spreng et al., 1996).

8.3   Consumption Processing Model

Several models to assess consumption satisfaction have been developed. As noted earlier, the majority of these approaches suggest that consumption satisfaction is a relative concept and judged in relation to a standard. While several comparison standards have been proposed in the literature, no consensus exists concerning which standard might be the most appropriate, that is, which standard best predicts consumption satisfaction (Cote et al., 1989).

To address, in part, the limitations of the Expectancy-Disconfirmation approach and include components of other published models, Oliver (1997) added an appraisal framework and proposed a general consumption processing model (Fig. 8.2).

To summarize the model, the sequence above the "consumption–satisfaction/dissatisfaction" link shows the non-processing phase of consumption, whereby consumers react to the consumption outcomes with more or less spontaneous effect. This unapprised pleasure/displeasure response is perhaps typically reflected in traditional hedonic measures of foods evaluated under blinding conditions.

The lower part of the model refers to the processing sequence which begins with the Expectancy-Disconfirmation model and includes an attribution processing component and an affect component.

Both the non-processing and processing sequences then affect the satisfaction construct. Other influences include anhedonic cognition, that is, unemotional observations regarding a product, as represented by the direct link between outcomes and satisfaction.

To illustrate how the consumption processing model could be deployed in a practical research sense, Oliver suggested a 12-item consumer satisfaction scale in which each scale item relates to some dimension (eg, expectations, performance, disconfirmation, attribution) of the general satisfaction model.

As in the example noted below, in addition to the satisfaction anchor, the scale consists of two overall performance or purchase evaluation items, including one which can be viewed as a proxy for quality, three attribution items, two affect items, a need fulfillment item, a dissonance item, a regret item, and a negative disconfirmation item.

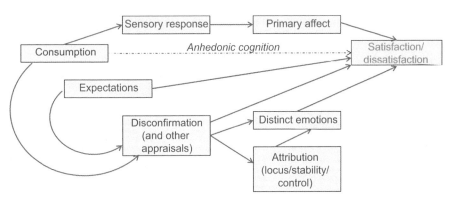

Figure 8.2 General consumption processing model.
Adapted from Oliver, R.L., 1997. Satisfaction. A Behavioral Perspective on the Consumer. McGraw-Hill.

Oliver notes that the scale can be adapted to reflect the focus of the researcher, in particular noting the inclusion of positive disconfirmation (met expectations) within the scale if this is not separately covered within a study, or equity in terms of a perceived "fair deal" if this is required.

8.4   Case Study: Citrus-Flavored Green Teas

Oliver's published application of the scale across a range of markets show consistently good reliability with scores of 0.75–0.9+. With this in mind, Campden BRI deployed an adapted version of the scale in a trial on chilled green teas.

8.4.1   Materials and Methods

In this study, five chilled, citrus-flavored green teas (see Fig. 8.3) were assessed by five different sets of a minimum of 60 pre-recruited respondents in a central location test format. Multiple assessment sessions were convened over a 2-day period, with each session lasting around 45 min and attended by about 25 respondents per session. The sessions were convened in one location in the United Kingdom.

To qualify to complete the study, all respondents were regular buyers and consumers of non-carbonated soft drinks, and all were non-rejectors of chilled green teas and of the flavors under test. Other standard market research recruitment criteria were applied.

The cleaned datasets of 328 respondents in total were used in the analysis. Thirty-eight percent of qualifying respondents were male; 36% were aged 18–30 years; 36% 31–45 years, and 29% 46–75 years. Twenty percent of respondents were of socio-economic grade AB, 54% C1, and 26% C2.

The five sets of consumers were similar in terms of age, gender, frequency of purchase, self-declared health orientation, and product familiarity.

The teas selected were, at the time of writing, all currently available in the UK market and were comparable from a quality/price perspective; however, the packaging design, information, and physical characteristics varied.

The test protocol consisted of two parts. In part A, each respondent was asked to complete a choice task where they had to select from the five products, which one they

Figure 8.3 Chilled green tea products included in the case study.

would be likely to buy, and note the reasons for their selection. The product selection was recorded, and respondents went on to complete part B of the study.

In part B, each respondent completed a taste task, evaluating first a dummy sample (Tynbark Green Tea Lemon Drink), followed by a blind hedonic assessment of the product selected in part A, then an informed hedonic assessment of the selected product where the packaging and product were presented simultaneously.

The measurement of consumption satisfaction was undertaken following the informed assessment stage using an adapted scale (see Table 8.3). A number of other measures such as perceived sensory attributes and feeling post-consumption were also captured.

A 9-point hedonic scale was used to capture respondents' degree of liking. A 7-point Likert scale (Agree strongly/Disagree strongly) was used to measure respondents' agreement with the consumption satisfaction scale items, the order of which was rotated among respondents. A Check-All-That-Apply (CATA) approach was used to capture associations between attributes and samples in relation to packaging, and expected and perceived sensory attributes, and a 5-point scale was used to measure purchase intent.

8.4.2   Statistical Analysis

Data were captured via laptops using Compusense at hand sensory software. Once exported, the data were analyzed utilizing a range of software including Minitab v17.2.1, R v3.2.0, SPSS v22, and XLSTAT v2015.1.03.

A range of statistical tests were applied to the data: *t*-tests when comparing two conditions (blind versus informed) for differences in mean scores (run at 5% significance); Kruskal–Wallis nonparametric alternative to analysis of variance (ANOVA) followed by post hoc tests using Dunn's procedure (both at 5% significance) when comparing three and more products or conditions for differences in mean scores; correspondence

Table 8.3 **Chilled Green Tea: Items in the Consumption Satisfaction Scale**

1. This was one of the best chilled green teas I could have bought *(overall performance and quality)*.
2. This chilled green tea was exactly what I needed *(need fulfillment)*.
3. **This chilled green tea did not taste as good as I thought it would** *(negative disconfirmation)*.
4. I am satisfied with my decision to choose this chilled green tea *(satisfaction anchor)*.
5. **Now that I have drunk that chilled green tea, I have mixed feelings about choosing it again** *(cognitive dissonance)*.
6. My choice to choose this chilled green tea was a wise one *(success attribution)*.
7. **If I were to choose a chilled green tea again, I would choose a different product** *(regret)*.
8. I truly enjoyed this chilled green tea *(positive affect)*.
9. **I am not happy that I chose this chilled green tea** *(negative affect)*.
10. Drinking this chilled green tea has been a good experience *(choice evaluation)*.
11. I did the right thing choosing this chilled green tea *(success attribution)*.
12. This chilled green tea exceeded my expectations *(positive disconfirmation)*.

analysis of CATA frequency data, chi-square tests (at 5% significance) followed by post hoc Marascuilo procedure; and principal components analysis to summarize the product differences and relationships between the different conditions (expected, actual blind, actual informed).

Cronbach's alpha coefficient was run to assess the reliability of the satisfaction scale.

8.4.3 Results

Respondents were asked to indicate from a prompted list of 14 statements those that best described the reasons for their product selection.

The varying packaging design, information, and physical characteristics were reflected in reasons for selection with, for example, "appealing flavor combination" being cited as the principal reason for selection of one product, while "packaging attractiveness" was cited as the principal reason for some of the other products.

So, while the demographic profile characteristics were similar between the five sets of respondents, the reasons for product selection varied.

In addition, perceived packaging attributes (22 in total) were also captured through a CATA approach along with the expected sensory attributes (13 in total) as conveyed by the packaging, shown in the following correspondence analysis maps (Lê and Worch, 2015), Figs. 8.4 and 8.5.

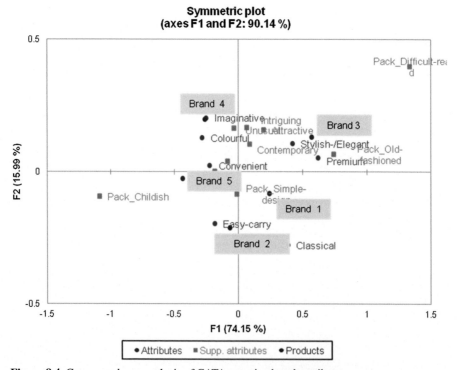

Figure 8.4 Correspondence analysis of CATA perceived pack attributes.

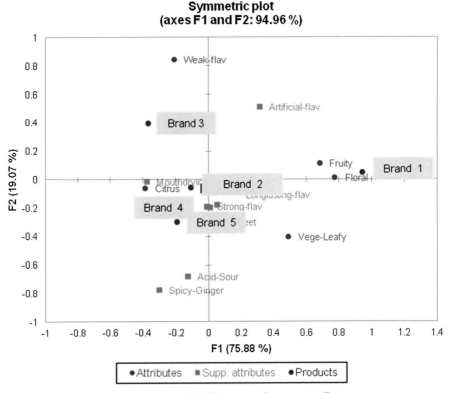

Figure 8.5 Correspondence analysis of CATA expected sensory attributes.

(The attributes included in the CATA questionnaire were generated by a trained sensory descriptive panel and checked for consumer interpretation.)

Perceived differences in packaging associations can be seen with, for example, Brand 3 being associated with "stylish/elegant" and "premium" attributes, and Brand 5 with more "classical" attributes.

The packaging also conveyed differences in the expected sensory qualities of the products. For example, Brand 3 was anticipated to be more "fruity/floral" in character, while Brand 4 was anticipated to be particularly "citrusy."

In terms of packaging liking, all samples were liked moderately or very much (which is not altogether surprising as each respondent had selected their preferred product), although Brand 3 was liked the most and Brand 5 the least (Fig. 8.6).

In terms of overall liking, all products were liked more in the informed condition compared to the blind condition. For Brands 1, 2, and 3, the packaging and informed liking measures were rated consistently and significantly higher than blind overall liking; for Brands 4 and 5, packaging liking was significantly higher than informed taste liking, which was significantly higher than blind taste liking.

In terms of the satisfaction statements, the mean scores reflected the rating for overall liking, with two groups of products being identified. In the majority of cases,

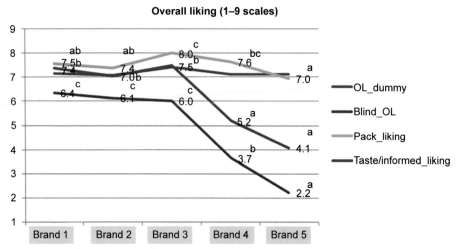

Figure 8.6 Overall hedonic liking of dummy sample, products in blind condition, products in informed condition, and pack liking. Kruskal–Wallis tests for comparing more than two independent samples and multiple pairwise comparison tests using Dunn's procedure (two-tailed tests) were run. Different letters indicate significant differences at 5% between products within each condition.

the statement scores for Brands 1, 2, and 4 were significantly different from Brands 3 and 5 for both the positive and negative statements (Fig. 8.7).

A summarized satisfaction score (ie, the mean rating across all the positive attributes less the mean rating across all the negative attributes) showed a similar trend to the overall liking score recorded under both the blind and informed conditions (see Fig. 8.8).

While the post hoc multiple comparison test could divide the samples into two product groups, liked and disliked across the attributes, product differences within the two groups were not statistically differentiated at 5% significance in this trial.

So what potential additional insight can an assessment of consumer satisfaction add to our understanding of the products over and above that obtained from typical blind/informed hedonic measurements?

If we relate the satisfaction statements back to the constructs covered in the model, perhaps a further sense of how respondents viewed the products emerges (Fig. 8.9).

It is encouraging that the liked and disliked products were consistently differentiated across the satisfaction measures. For example, the least liked samples were more strongly associated with the negative constructs of "regret" and "cognitive dissonance," falling short of meeting needs or expectations, while the reverse was true for the more liked products.

However, additional valuable insight would have been gained if the scale had revealed greater discrimination between products in the most liked and least liked groups.

It could, of course, be that respondents considered that there was no difference between the samples within these two groups. However, it should be noted that in this trial the consumption satisfaction scale was deployed within a central location test

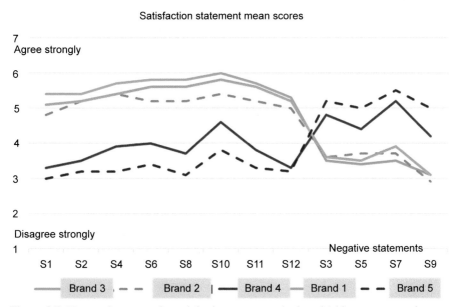

Figure 8.7 Mean rating scores for satisfaction statements by brand (chi-square test and post hoc Marascuilo procedure were run at 5% significance on agreement scales).

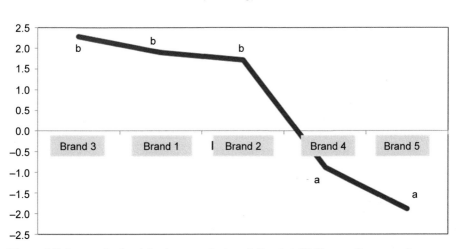

Figure 8.8 Summarized satisfaction score by brand. Kruskal–Wallis tests for comparing more than two independent samples and multiple pairwise comparison tests using Dunn's procedure (two-tailed tests) were run. Letters indicate significant differences ($p < 0.05$).

format, which may have suppressed consumers' response to some of the scale dimensions, given consumers' limited exposure to the product.

It is possible that longer term exposure to the products, through, for example, a home placement test, may have led to some of the "within" product group differences

Figure 8.9 Mean scores for satisfaction statements by model component. 1. Quality evaluation, 2. Need fulfillment, 3. Negative confirmation, 4. Satisfaction anchor, 5. Cognitive dissonance, 6. Success attribution, 7. Regret, 8. Positive affect, 9. Negative affect, 10. Positive affect, 11. Success attribution, and 12. Positive confirmation.

being discriminated more clearly, providing those involved in packaging and product development a deeper understanding of consumer engagement with the product.

In addition to the limitations of the test format in this trial, the reliability of the satisfaction scale linked to the interrelatedness of the items and the extent to which they measured the same construct may also have led to a lack of product discrimination.

Scale reliability was examined using Cronbach's alpha coefficient. In this trial, the coefficient based on standardized items had a value of 0.65. Different reports suggest acceptable values of alpha ranging from 0.70 to 0.95 (DeVellis, 2003).

The coefficient value increased to 0.97 when the negative items were removed. This, and a review of the inter-item correlation matrix, suggests that perhaps respondents in this trial may have had some difficulty in responding to the negative statements (although this is not an untypical response in consumer studies).

In addition, the statement "Drinking this chilled green tea has been a good experience *(choice evaluation)*" showed a lower inter-item correlation than the other positive statements. Responses here may have reflected ambiguity in interpretation of the statement in that respondents may have seen this as relating to the whole test experience rather than specifically the product itself.

Improvements to the wording of some of the scale items may have led to an improvement in product discrimination in this trial.

8.4.4 Delight

The construct of "delight" has been growing in importance in the satisfaction research literature (Burns and Evans, 2000).

In psychology, delight is regarded as a second-order emotion which specifically results from a combination of pleasure and surprise. In this trial, we aimed to assess the "surprise" quantity through measuring respondents' perceptions of perceived uniqueness (along with other measures) in respect of the packaging and also of the product after consumption.

As reflected in the principal components analysis (see Fig. 8.10), among the three most liked products, Brand 3 was considered more "unique," particularly in relation to its packaging (statistically significant at 10%), compared to Brand 2.

The correspondence analysis shown earlier illustrated that Brand 3 packaging was associated with "stylish/premium" qualities, and it was the most liked in terms of

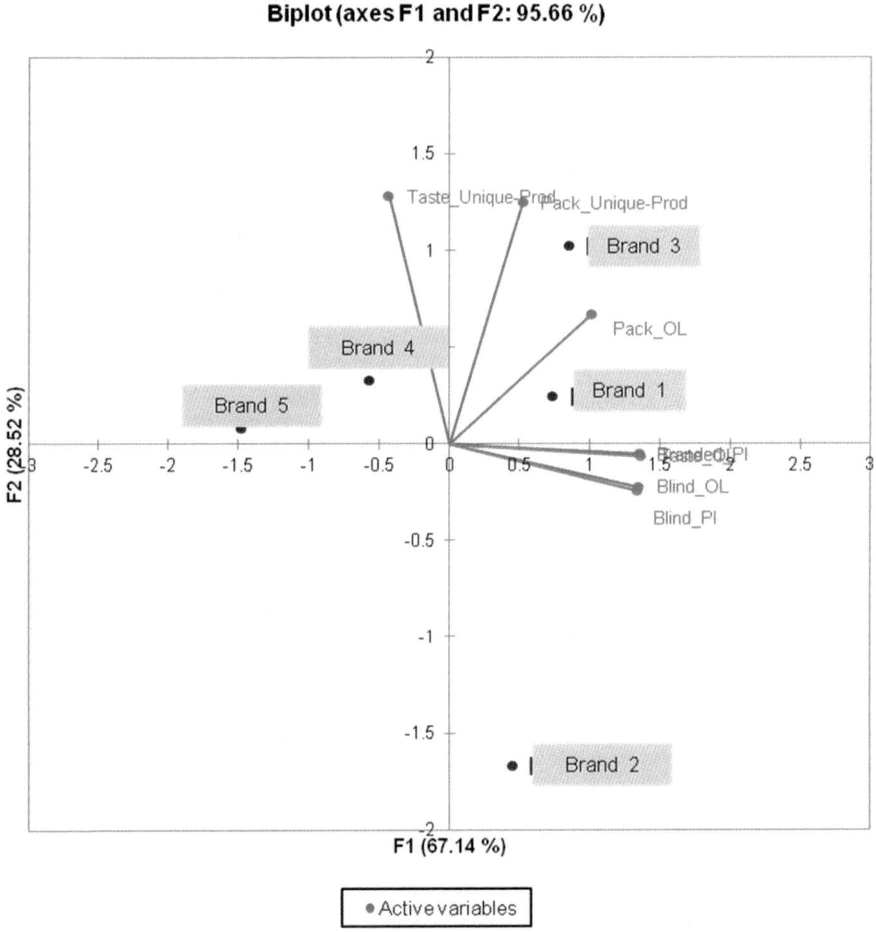

Figure 8.10 Principal component analysis on all liking variables, purchase intent, and perceived informed flavor and pack uniqueness.

packaging; a comparison of overall liking under blind and informed conditions suggests that there may well have been an assimilation effect.

The measure of uniqueness may therefore be capturing an additional delight (surprising pleasure) component of satisfaction which could explain the propensity for Brand 3 to be rated indicatively higher than the other liked brands across the positive satisfaction statements.

While the phrasing of the "uniqueness" measure included in the study may be open to improvement, the inclusion of an attribute in a similar vein to this within the satisfaction scale would appear to be a valuable addition that may contribute to an enhanced discrimination between products.

8.4.5 Segmentation

The results from this trial were considered at an aggregate level only, as the limited respondent sample size per product assessment restricted a meaningful statistical segmentation of the data.

However, in an exploratory study undertaken by Campden BRI (Chen, 2014), different consumer groups were identified through Ward's Hierarchical Clustering of responses to a similarly constructed consumption satisfaction scale to the chilled green tea study. In this investigative study, a three-cluster solution was selected, with the groups labeled as Delighted, Satisfied, and Dissatisfied.

A one-way ANOVA followed by Tukey's multiple comparison test at 5% significance identified that the Delighted group significantly recorded higher mean scores on the positive statements, demonstrating a highly satisfactory integrated product and packaging experience. In contrast, the Dissatisfied scored highly on the negative statements.

The different levels of satisfaction were found to be aligned with respondents' willingness to repurchase the products, where delighted consumers were most likely to purchase the product again, while the dissatisfied consumers were least likely to repurchase the product.

A segmentation of consumer responses based on the satisfaction scale may offer further insight and discrimination between products that are similarly rated on a traditional hedonic rating scale.

8.5 Conclusions

As noted earlier, many product categories are characterized by large numbers of products competing with very similar specifications and features. Benchmarking and matching the best in class is no longer sufficient to win new customers, and products that merely satisfy fail to encourage loyalty in existing customers.

Numerous commentators have therefore discussed the requirement to move beyond simply satisfying consumers, captured in part through measures such as hedonic liking, to delighting them. Engaging the consumer at an emotional as well as functional level is now regarded as key to the long-term success of a product.

This had led to a growing interest in investigating approaches that capture consumer responses to products that go beyond traditional sensory and liking measures. These approaches include, for example, assessing emotional responses (King and Meiselman, 2010), temporal drivers of liking (Thomas et al., 2015), as well as considering the effect of situation and context in the evaluation of food and beverage products (Petit and Sieffermann, 2007).

For fast-moving consumer goods (FMCG) in particular, packaging can be considered as a contributor to context, with recent cross-modal research (Velasco et al., 2014) illustrating the potential effect on consumers' dissatisfaction or delight with a product that can be evoked where there is dissonance between the packaging and the product experience.

Indeed, Piqueras-Fiszman and Spence (2015) provide a comprehensive review on the effects of expectations on the sensory perception of foods. This review highlights how language, pictures, names, images, and/or symbols can all give rise to different expectations and reactions to a food from a consumer perspective.

Traditional consumer sensory hedonic measures, linked with other data such as that obtained from trained sensory panels, have been valuable in optimizing product design at a functional level. These models are however limited in capturing a more complete consumer response, particularly when packaging-evoked expectations are taken into consideration.

Understanding consumer response to the integrated product and packaging experience in terms of the various dimensions of satisfaction (disconfirmation, attribution, affect, and need fulfillment) is an invaluable step for brand owners. Indeed, within the consumer and sensory science discipline, there is growing interest in the application of consumption satisfaction and the insight that this approach may provide in obtaining a more holistic characterization of consumers' response to products.

In obtaining this broader perspective on consumption satisfaction, the brand owner has the potential to initiate appropriate development strategies that will secure a strong preference for the brand, a commitment to continued purchase, and the foundation for long-term brand survival.

As noted, the satisfaction response is a complex construct consisting of cognitive (processing states), hedonic, and affective (emotions) components. The satisfaction scale reported in this chapter has been deployed across numerous categories, and although it has received little attention to-date within the food and beverage sector, it may offer a useful framework for further investigation.

References

Burns, A.D., Evans, S., 2000. Insights into customer delight (Chapter 29). In: Scrivener, S.A.R., Ball, L.J., Woodcock, A. (Eds.), Collaborative Design. Springer, London.

Bitner, M.J., 1990. Evaluating service encounters: the effect of physical surroundings and employee responses. Journal of Marketing 54, 69–82.

Cardello, A.V., 2007. Measuring consumer expectations to improve food product development. In: MacFie, H. (Ed.), Consumer-led Food Product Development. Woodhead.

Chen, X.M., 2014. Assessing Dissonance Between Packaging and Product Experience: A White Wine Study Report No 379. Campden BRI R&D.

Cote, J.A., Foxman, E.R., Cutler, B.D., 1989. Selecting an appropriate standard of comparison for post-purchase evaluations. Advances in Consumer Research 16, 502–506.

Dannemillar, K.D., Jacobs, R.W., 1992. Changing the way organisations change: a revolution of common sense. The Journal of Applied Behavioural Science 28 (4), 480–498.

DeVellis, R., 2003. Scale Development: Theory and Applications. Sage, Thousand Oaks, CA.

Festinger, L., 1957. A Theory of Cognitive Dissonance. Stanford University Press, Stanford, CA.

Giese, J.L., Cote, J.A., 2000. Defining consumer satisfaction. Academy of Marketing Science Review. 1. http://www.amsreview.org/articles/giese01-2000.pdf.

Gilovich, T., Medvec, V.H., 1995. The experience of regret: what, when and why. Psychological Review 102, 379–395.

Gordon, W., 2011. Behavioural economics and qualitative research – a marriage made in heaven? International Journal of Market Research 53 (2), 171–185.

Halstead, D., Hartman, D., Schmidt, L.S., 1994. Multi source effects on the satisfaction formation process. Journal of the Academy of Marketing Science 22 (2), 114–129.

Kahneman, D., Tversky, A., 1979. Prospect theory: an analysis of decisions under risk. Econometrica 47, 263–291.

Kahneman, D., 2003. A psychological perspective on economics. American Economic Review 93 (2), 162–168.

King, S.C., Meiselman, H.L., 2010. Development of a method to measure consumer emotions associated with foods. Food Quality and Preference 21, 168–177.

Köster, E.P., 2009. Diversity in the determinants of food choice: a psychological perspective. Food Quality and Preference 20, 70–82.

Lê, S., Worch, T., 2015. Chapter 4 in Analysing Sensory Data in R. CRC Press.

Oliver, R.L., 1997. Satisfaction. A Behavioural Perspective on the Consumer. McGraw-Hill.

Petit, C., Sieffermann, J.M., 2007. Testing consumer preferences for iced-coffee: does the drinking environment have any influence? Food Quality and Preference 18, 161–172.

Piqueras-Fiszman, B., Spence, C., 2015. Sensory expectations based on product-extrinsic food cues: an interdisciplinary review of the empirical evidence and theoretical accounts. Food Quality and Preference 40, 165–179.

Samuelson, W., Zeckhauser, R., 1988. Status quo bias in decision making. Journal of Risk and Uncertainty 1, 7–59.

Schifferstein, H.N.J., Kole, A.P.W., Mojet, J., 1999. Aysmmetry in the disconfirmation of expectations for natural yogurt. Appetite, 32, 307–329.

Schwartz, B., Ward, A., 2004. Doing Better but Feeling Worse: The Paradox of Choice. Positive Psychology in Practice. Wiley.

Spreng, R.A., Mackenzie, S.B., Olshavsky, R.W., 1996. A re-examination of the determinants of consumer satisfaction. Journal of Marketing 60, 15–32.

Thomas, A., Visalli, M., Cordelle, S., Schlich, P., 2015. Temporal drivers of liking. Food Quality and Preference 40, 365–375.

Velasco, C., Woods, A.T., Deroy, O., Spence, C., 2014. Hedonic mediation of the cross modal correspondence between taste and shape. Food Quality and Preference 41, 151–158.

Westbrook, R.A., Reilly, M.D., 1983. Value–percept disparity: an alternative to the disconfirmation of expectations theory of customer satisfaction. In: Bogozzi, P.R., Tybouts, A. (Eds.), Advances in Consumer Research, vol 10. Association for Consumer Research, Ann Arbor, MI, pp. 256–261.

Yeomans, M., Chambers, L., Blumenthal, H., Blake, A., 2008. The role of expectancy in sensory and hedonic evaluation: the case of smoked salmon ice-cream. Food Quality and Preference 19, 565–573.

Further Reading

Schneider, J., Hall, J., 2011. Why Most Product Launches Fail. Harvard Business Review. https://hbr.org/2011/04/why-most-product-launches-fail.
Schifferstein, H.N.J., 2001. Effects of product beliefs on product perception and liking. In: Frewer, L., Risvik, E., Schifferstein, H. (Eds.), Food, People and Society: A European Perspective of Consumers Food Choices. Springer Verlag, Berlin, pp. 73–96.

Looking Forward

The content of this book reflects an emerging and growing focus on the interrelationship between packaging and product, particularly in relation to the fast-moving consumer goods (FMCG) sector. This interest is being shaped principally by brand owners as they seek to optimize the congruency between expectations and the consumption experience.

At a tactical level, aiming for this congruency is about ensuring that consumers remain satisfied with the product and hopefully continue to repurchase. However, at a strategic level, this approach opens up potential routes and ways of thinking about product development and innovation that can have significant implications for a brand and its positioning in the market.

The contributing authors to this book have drawn conclusions with respect to each of their specific chapters. It is therefore not intended that this chapter replicates or adds conclusions to those noted, but rather looking forward, it considers the implications of the preceding chapters on product and packaging innovation in the context of the emerging discipline of sensory branding, an area that is receiving increasing focus in both academia and industry.

From the academic literature, two areas of study relevant to sensory branding are highlighted here, namely the "reward–pleasure cycle" and "cross modality."

In the 2015 book *The Psychology of Desire* (Ed. Hofmann, W & Nordgren, L.F. Guilford Press), two psychologists, Kringelbach & Berridge, introduced the concept of the "reward–pleasure cycle." According to this model, rewards are described as "motivational magnets" that initiate, sustain, or switch states, or in an FMCG context to trial, repeat buy or change brands.

Typically, rewarding moments go through a phase of wanting (ie, expectations) a reward and liking (ie, consumption experience) of the reward, both of which contribute to learning which help form the associations, representations, and predictions about future rewards based on past experiences (ie, satisfaction and brand loyalty).

If we consider that packaging can, among other things, be a key driver in shaping consumer wanting of the product, that consumption determines product liking, and that expectations coupled with the consumption experience inform satisfaction (and subsequent related stages of brand loyalty), parallels with models such as that discussed by Kringelbach & Berridge are evident.

Interestingly, Kringelbach & Berridge in their lab have gone on to demonstrate that wanting and liking each have distinguishable neurobiological mechanisms. Wanting is characterized by a reward/motivation system, while liking is characterized by a sensory pleasure in-the-moment/experiential system.

While these two systems can operate separately in terms of developing a strong product and packaging proposition, and thereby the chances of long-term consumer loyalty, it is necessary to understand both the motivations that propel consumers to make certain choices that look at delivering the fulfillment they seek, and the emotional representations associated with the experiential system.

As outlined by Forbes ("Toward a Unified Model of Human Motivation." *Review of General Psychology*. Vol. 15, No. 2. 2011), motives are limited to a universal set of nine (security, identity, mastery, empowerment, engagement, achievement, belonging, nurture, and esteem), while emotional representations can be more diverse.

Both of these systems (the reward/motivation and sensory pleasure in-the-moment/experiential system) are characterized by what behavioral economists refer to as system 1 thinking—ie, thinking that it is rapid, associative, intuitive, and nonconscious. It is important to reiterate, as noted in previous chapters, that these intuitive processes deal with the large majority of problems and choice decisions on a day-to-day basis.

Sensory branding approaches can help organizations develop brands and products that connect with consumers on a system 1, nonconscious level.

As noted by Simon Harrop (CEO of Brand Sense), a further key consideration in relation to sensory branding is that of cross modality—that is, how sensory signals from one sense can have an impact on and be interpreted through the others.

Harrop notes an example of the way in which the constituent parts of the sensory system are interrelated by reference to the work undertaken by the neuroscientist Cyriel Pennartz published in "Identification and integration of sensory modalities: Neural basis and relation to consciousness. *Consciousness and Cognition 18*. 2009."

In this paper, Pennartz provides a graphical representation of the estimated correlation strengths between the sensory modalities. In this representation, the visual domain is strongly connected to the tactile, proprioceptive, vestibular, and auditory modalities. Weaker relationships are present between vision on the one hand, and olfaction and taste on the other. (Additional modalities associated with the somatosensory system such as thermo and nociception are not included in the representation.)

What Harrop and others emphasize is that the cross modal utilization of the senses has to date been underplayed in much of brand and product development activity. However, psychological studies reveal that mono and cross modal signals from nonvisual senses can act as subconscious cues that strongly guide consumer behavior by linking emotional associations with certain actions, described as somatic markers by the neuroscientist Antonio Damasio.

A cross modal–based approach also addresses the limitations associated with engaging consumers on a mainly visual platform. Visual communication by definition connects with consumers on a conscious level, as images are largely processed actively. However, this form of communication is vulnerable to being filtered out unconsciously by consumers, given the quantity of visual stimuli people are bombarded with in their everyday lives.

So understanding the underlying category motivations, the desired emotions of the moment, and utilization of cross-modal interactions can provide product development and brand teams with opportunities to connect with their target audience at deep motivational and emotional levels that differentiate a brand from its competitors, are highly relevant to the way consumers make decisions, and are relatively difficult for competitors to emulate.

Understanding these elements reiterates the necessity of deploying some of the methods and techniques for obtaining consumer insights in these areas, as discussed in the preceding chapters.

Sensory branding is one evolving approach that brand owners can apply to develop these deep motivational and emotional connections. Sensory branding is about creating new or emphasizing sensations; ie, the touch, taste, smell, sound, and look of a product that affect consumer emotions, memories, perceptions, preferences, and choices.

This approach not only helps brand owners optimize the integration of the packaging and product experience but also guides the tone and style of brand development and marketing activities more generally.

For example, when it comes to understanding "wanting" (expectations), a sensory branding and product development perspective can be taken when considering questions such as:

- How do consumers differentiate between brands in this market?
- How is our brand perceived by consumers?
- What is the opportunity that we might want to pursue?
- Who is the customer who we want to target?
- What benefits does the proposed product provide to solve consumer problems?

Additionally, a sensory branding perspective can be taken when it comes to understanding product "liking" (experience) and considering questions such as:

- Which features create a unique sensory signature for our brand?
- Which prospective features are worth investing in?
- What shall we actually include in the final product?
- How much will people be prepared to pay for the product?

And finally, a sensory branding perspective can be taken when it comes to understanding the "learning" (outcomes) and associated questions such as:

- To what extent does the product meet consumer needs?
- Will the consumer repurchase the product?
- Which features support the brand proposition?
- What areas need to be adjusted to enhance consumer engagement with the product?

These questions are typically addressed through a systematic approach to creating a brand sensory signature and system.

A sensory branding approach typically consists of four broad stages. The first stage involves undertaking a brand audit to assess an organization's market and market positioning. Practically, this involves both workshops with internal teams and qualitative research with a brand's consumer audience, with the aim of crafting a desired positioning and a set of related image attributes.

The next stage involves translating this desired positioning into a sensory expression of the brand and thereby creating emotional, functional, and sensory specifications for the associated product, packaging, and going further, brand and marketing communications. Again, workshops with internal teams and third parties such as flavor houses, packaging designers and technologists, etc. are used to develop these specifications, which are then used to produce multisensory prototypes that aim to reflect the desired positioning.

The third stage involves evaluating the multisensory prototypes in terms of consumer needs, expectations, experiences, and satisfaction. The aim here is to establish the degree of congruency between the outputs from stages 2 and 3. Further refinement of the product and packaging may be required to ensure alignment with the desired brand values.

The final step is that of implementation. This stage comprises an internal focus, namely cascading the sensory brand signature through various teams across the organization involved in the delivery of the consumer experience. This clearly has implications for organizational design in terms of operating groups and implementation teams.

There is also an external focus to this stage, namely that of tracking the performance of the multisensory product and packaging proposition in market. Importantly, this activity will inform business metrics around the effectiveness of the sensory branding initiative, reflected in measures such as brand value and return on investment.

In summary, there is growing interest in the potential that a cohesive sensory brand identity system can offer in terms of enabling brands to engage with consumers in a differentiated way. Many of the issues and opportunities for consideration when it comes to product and packaging design and the methods and techniques for assessing the integrated product and packaging consumer experience discussed in this book represent valuable inputs to the sensory branding process.

It is hoped that the individual chapters and the book as a whole have stimulated the reader to explore how a multisensory approach in creating an integrated packaging and product consumer experience can offer significant opportunities for sustainable brand development.

Peter Burgess
Campden BRI
January 2016

Index

'*Note*: Page numbers followed by "f" indicate figures, "t" indicate tables.'

A

Affective priming, 126
Associative memory networks, 126–128
Auditory packaging design, 12
Augmented reality (AR), 61–62
Autonomic nervous system (ANS)
  autonomic measures, 89, 90f
  eye tracking, 90–91
  facial electromyography (fEMG),
    89–90
  galvanic skin response (GSR), 90
  heart rate variability (HRV), 90

B

Behavior coding, 98–99
Best-Worst scaling, 151–153
Biscuit packaging, 49–50, 50f
Bivalent satisfiers, 168
Blood oxygen level-dependent
  (BOLD), 94

C

Cambridge Simulation Glasses, 46, 46f
Cambridge Simulation Gloves, 47–48, 47f
Capacitive coupling, 73
CEN15945 Technical Specification, 43–44
Check-All-That-Apply (CATA) approach,
  171
Citrus-flavored green teas, 170, 176f
  correspondence analysis, 172, 172f–173f
  delight, 177–178, 177f
  materials and methods, 170–171, 171t
  mean rating scores, 173–174, 175f
  segmentation, 178
  statistical analysis, 171–172
  summarized satisfaction score, 174, 175f
Cognitive psychology, 129–130
Conjoint analysis, 149–151
Consumer choices, 161–163
Consumers' mindset, 174f

citrus-flavored green teas, 170, 176f
  correspondence analysis, 172,
    172f–173f
  delight, 177–178, 177f
  materials and methods, 170–171, 171t
  mean rating scores, 173–174, 175f
  segmentation, 178
  statistical analysis, 171–172
  summarized satisfaction score, 174,
    175f
consumer choices, 161–163
consumption processing model, 169–170,
  169f
expectations and satisfaction, 165t
  affect, 168
  assimilation-contrast approach, 163,
    164f
  Expectancy-Disconfirmation model,
    163
  generalized consumption satisfaction
    scale, 165, 166t
  Gleicher change formula, 163
  need fulfillment, 168
  purchase apprehension, 166–167
  purchase attribution, 167
Consumption processing model, 169–170,
  169f
Covent Garden Co.'s Tetra Pak, 7–8, 8f
Create phase, 43
Customer order fulfillment options,
  69, 70f

D

Dexterity impairment simulators,
  47–48, 47f

E

Ease of opening, 10–11, 11f
Electroencephalography (EEG), 92–93
EmoSemio approach, 83

Emotion measurement methods
  autonomic nervous system (ANS)
    autonomic measures, 89, 90f
    eye tracking, 90–91
    facial electromyography (fEMG),
      89–90
    galvanic skin response (GSR), 90
    heart rate variability (HRV), 90
  from brain
    brainwave categories, 91–92
    electroencephalography, 92–93
    neuroimaging, 93–95
  characterization, 78
  conceptualization, 78
  current research interests, 78–79, 79t
  EmoSemio approach, 83
  emotive projective test (EPT), 89
  EsSense Profile™, 81–82
  and expectations, 106–107, 108f
  Geneva Emotions and Odor Scale
    (GEOS), 81–82
  Implicit Association Test (IAT), 88
  neuroscience, 80
  predetermined questionnaires, 82
  product emotion measuring instrument,
    87–88
  product experience
    brand-first strategy, 100
    food emotions, 99
    liking and choice, 100–101
    product-first strategy, 100
    sensory and branding, 101–103
    sensory drivers of emotions, 100–101
  reducing ambiguity, 84
  Self Assessment Manikin (SAM)
    technique, 85–87, 85f
  standardized questionnaire, 82
  through behavior, 95–96
    behavior coding, 98–99
    facial coding, 96–98, 96f
  unbranded and branded food products, 104t
    blind and informed conditions, 104, 105t
    EmoSemio questionnaire, 103
    experimental design, 103, 103f
    happy memories, 105–106
  using questionnaires, 84–85
  verbal emotional approach, 81
  visual self-report, 85
Emotive projective test (EPT), 89

Empathy tools, 44–45
EsSense Profile™, 81–82
Evaluate phase, 43–44
Expectancy-Disconfirmation model, 163, 166
Explicit codes, 98–99
Explicit methods
  conjoint analysis, 149–151
  holistic approaches, 153–155
  large-scale quantitative assessment
    individual elements, 140
    multisensory experience, 140
    9-point category scales, 142
    5-point JAR scales, 142
    preference mapping, 141
    product-concept fit, 141
  MaxDiff, 151–153
  packaging in-home and in-store,
    consumers' interaction, 148–149
  small-scale exploration of consumers'
    attitudes
    enabling techniques, 144
    projective techniques, 144
    sentence completion, 145
    tactile assessment, 146–147
    word association, 144–145
Explore phase, 41–43, 42f

F
Facial active coding system (FACS), 96
Facial coding, 96–98, 96f
Facial electromyography (fEMG), 89–90
Food labels, 23–24
Front-of-pack (FOP) messages, 25
Functional magnetic resonance imaging or
  functional MRI (fMRI), 94
Functional near-infrared spectroscopy
  (fNIRS), 95

G
Galvanic skin response (GSR), 90
Geneva Emotions and Odor Scale (GEOS),
  81–82

H
Health/nutrition claims, 29–31
Heart rate variability (HRV), 90
Hierarchical multivariate factor analysis
  (HMFA), 155

I

Impairment Simulation Software, 48, 49f
Implicit Association Test (IAT), 88
Implicit codes, 98–99
Implicit reaction-time (IRT) tests, 125–126
Inclusive design process, 40
  Create phase, 43
  easy-open pasta packaging, 55, 55f
  empathy tools, 44–45
  Evaluate phase, 43–44
  Explore phase, 41–43, 42f
  Manage phase, 44
  model for, 41, 41f
  simulation methods, 45
    biscuit packaging, 49–50
    resealable labels, 51–52
    software simulators, 48–49
    wearable simulators, 45–48

L

Last-mile logistics, 69

M

Manage phase, 44
MaxDiff, 151–153
Monovalent dissatisfiers, 168
Monovalent satisfiers, 168
Multisensory packaging design
  auditory packaging design, 12
  ease of opening, 10–11, 11f
  individual/cultural differences, 13
  neuroscience-inspired packaging design
    electrophysiological brain-imaging
      techniques, 3
    implicit association test and
      eye-tracking, 2
    online testing platforms, 3
  packaging color, 4
    Barilla pasta's blue, 6
    Cadbury's Dairy Milk, 6
    crisps, 4–5
    Heinz brand color, 6
    Walkers brand, 4–5
  packaging shape
    Coca-Cola contour bottle, 7
    Covent Garden Co.'s Tetra Pak, 7–8, 8f
    crossmodal correspondences, 8
    Wishbone salad dressing bottle, 6–7, 7f
  packaging texture, 9–10
  packaging weight, 10
  tasty packaging, 13

N

Neuromarketing, 125
Neuroscience-inspired packaging design
  electrophysiological brain-imaging
    techniques, 3
  implicit association test and eye-tracking,
    2
  online testing platforms, 3
Neurosense/packaging
  American beauty products company
    background, 135
    conclusion, 136
    method, 135–136
    results, 136
  associative memory networks,
    126–128
  bottled wine packaging, 132
    conclusion, 134
    method, 132–133, 133f
    results, 133–134
  cognitive psychology, 129–130
  implicit reaction-time tests, 125–126
  implicit vs. explicit measures, 128–129
  neuroscience alternative, 124–125
  popular malt drink
    background, 130–131
    conclusions, 132
    method, 131
    results, 131–132
  product evaluation, 122–124
  self-report method, problems, 121–122
  System 1, 126–128
Noninclusive packaging
  easy-open packaging, 39
  failed to identify access/product
    information correctly, 38, 39f
  ingredients list, 38, 38f
  issues, package accessing, 38–39
  users struggling, open packaging and read
    cooking instructions, 37, 38f
Nutrition labeling, 28f
  attention, 29
  health symbols, 26–27
  mandatory information, 26–27
  motivation, 29

O

Omni-channel grocery retail
  Graze and Hello Fresh, 63
  issue of consumer trust, 64
  online shopping, 63
  top-up shops, 63–64
Omni-channel integrator, 72–73
Omni-channel retailing
  defined, 59
  omni-channel grocery retail
    Graze and Hello Fresh, 63
    issue of consumer trust, 64
    online shopping, 63
    top-up shops, 63–64
  packaging
    baked beans, 65–66
    customer satisfaction, 73–74
    identifying product variants, 67–68, 68f
    navigation route, 66
    omni-channel integrator, 72–73
    omni-channel transport, 69–72,
      70f–71f
    ready meal, visual presentation, 67, 67f
    supply chains, 65, 65f
    visual difference, 66, 66f
  product selections, 59
  shopping experience
    augmented reality (AR), 61–62
    mechanisms, 60–61
    number of channels, 60
    showrooming and webrooming,
      61, 62f
Omni-channel shopping experience
  augmented reality (AR), 61–62
  mechanisms, 60–61
  number of channels, 60
  showrooming and webrooming, 61, 62f
Omni-channel transport, 69–72, 70f–71f
On-pack educational messages, 32
  consumer processing, 25f
    front-of-pack (FOP) messages, 25
    hierarchy of effects, 24
    understanding and inferences, 25
  developments, 33
  food labels, 23–24
  types
    health and nutrition claims, 29–31
    nutrition labeling, 26–29
    sustainability information, 31–32

P

Packaging color, 4
  Barilla pasta's blue, 6
  Cadbury's Dairy Milk, 6
  crisps, 4–5
  Heinz brand color, 6
  Walkers brand, 4–5
Packaging shape
  Coca-Cola contour bottle, 7
  Covent Garden Co.'s Tetra Pak, 7–8, 8f
  crossmodal correspondences, 8
  Wishbone salad dressing bottle, 6–7, 7f
Packaging texture, 9–10
Packaging weight, 10
PAD. *See* Pleasure-Arousal-Dominance
      (PAD)
Personalised holistic approaches, 155
Personas
  black magic, 53, 54f
  life situations and capabilities, 52–53, 52f
  range of ages, 52–53, 52f
Pleasure-Arousal-Dominance (PAD),
      86, 86f
Preference mapping, 141
Product-concept fit, 141
Product emotion measuring instrument,
      87–88
Purchase apprehension, 166–167
Purchase attribution, 167

R

Rational consumer, 121
Resealable labels, 51–52, 51f

S

Self Assessment Manikin (SAM) technique,
      85, 85f
  consumer product testing, 86–87, 87f
  Pleasure-Arousal-Dominance (PAD), 86,
      86f
  two-dimensional mood mapping, 86–87,
      87f
Showrooming, 61, 62f
Simulation methods, 45
  biscuit packaging, 49–50
  resealable labels, 51–52
  software simulators, 48–49
  wearable simulators

cognitive ability, 45
dexterity impairment simulators,
   47–48, 47f
vision impairment simulators, 45–47, 46f
Skin conductive response (SCR),
   89–90
Software simulators, 48–49
Spreading activation, 127–128
Sustainability information, 31–32

T

Tasty packaging, 13

U

Ultra-flash profiling (UFP), 154

V

Vision impairment simulators, 45–47, 46f
Visual self-report, 85

W

Walmart, 71
Wearable simulators
   cognitive ability, 45
   dexterity impairment simulators, 47–48,
      47f
   vision impairment simulators,
      45–47, 46f
Webrooming, 61, 62f
WikiPearls, 13
Wishbone salad dressing bottle, 6–7, 7f

Printed in the United States
By Bookmasters